THEORY OF CONJECTURAL VARIATIONS

Series on Mathematical Economics and Game Theory

Series Editor: Tatsuro Ichiishi *(Ohio State University)*

Published

Vol. 2: Theory of Conjectural Variations
by C. Figuières, A. Jean-Marie, N. Quérou & M. Tidball

Series on Mathematical Economics and Game Theory

Vol. 2

THEORY OF CONJECTURAL VARIATIONS

Charles Figuières
University of Bristol, UK

Alain Jean-Marie
LIRMM, CNRS & University of Montpellier II, France

Nicolas Quérou
INRA-LAMETA & University of Montpellier II, France

Mabel Tidball
INRA-LAMETA, France

World Scientific

NEW JERSEY • LONDON • SINGAPORE • SHANGHAI • HONG KONG • TAIPEI • BANGALORE

Published by

World Scientific Publishing Co. Pte. Ltd.
5 Toh Tuck Link, Singapore 596224
USA office: Suite 202, 1060 Main Street, River Edge, NJ 07661
UK office: 57 Shelton Street, Covent Garden, London WC2H 9HE

British Library Cataloguing-in-Publication Data
A catalogue record for this book is available from the British Library.

THEORY OF CONJECTURAL VARIATIONS
Series on Mathematical Economics and Game Theory — Vol. 2

ISBN 981-238-736-6

Printed in Singapore by World Scientific Printers (S) Pte Ltd

To our families (C.F. and N.Q.),
to our parents (A.J.-M. and M.T.).

Preface

Game theory has proven useful in Economics, as a set of analytical tools to help us understand situations of interactions between agents promoting their partially conflicting interests. Two basic features of this theory of social interactions are:

(1) the fact that decision-makers are rational, *i.e.* they take decisions that are consistent with their own goals and the information they possess on their environment. Formally, they optimise well-defined exogenous objective functions, given a number of constraints;

(2) the efforts to explicit the way the decision-makers deal with the *strategic uncertainty* to solve their decision problems. In strategic situations, the benefit any agent can expect from her actions depends also on the actions of the other agents. The expression "strategic uncertainty" is meant to capture the fact that others' behaviours give to any individual decision problem some degrees of uncertainty, which optimists view as the zest of social life (pessimists would probably substitute the word plague for zest).

Equilibrium concepts single out specific outcomes of social interactions that are somehow consistent with the two features above. Game Theorists have retained alternative specifications for rationality and strategic uncertainty, in relation with the information each agent possesses, giving rise to different equilibrium concepts. The famous Nash equilibrium for instance is the outcome consistent with rational agents who take rival decisions as given when they optimise. As a second example, in a Stackelberg equilibrium there are two agents who take their decision sequentially; the first agent to move is referred to as the *leader*, whereas the second mover is called the *follower*. A Stackelberg equilibrium is an outcome consistent with the follower's rational behaviour given that she has observed the leader's move, and the leader's rational behaviour who can infer what will be the follower's rational reaction to her current decision.

Conjectural Variations Equilibria (CVE) have been introduced by two authors around 1930 as another possible solution concept in static games (Bowley (1924) and Frisch (1933)). According to this concept, agents behave as follows: each agent chooses her most favourable action taking into account that rival strategies are a conjectured function of her own strategy.

The concept of conjectural variations has been the subject of numerous theoretical controversies (see for instance Lindh (1992)). Nevertheless, economists have made extensive use of one form or the other of the CVE to predict the outcome of noncooperative behaviours in several fields of Economics. A considerable amount of empirical and econometric works also exists that evaluates what conjectures look like in particular game theoretic situations. More recently a renewal of interest for this concept has taken place using either a dynamic context, or situations of bounded (procedural) rationality, or both.

The purpose of this monograph is threefold. First, it is to present the concept of *conjectures* in games, and review the results of the literature on *conjectural variations* equilibria and their variants. Second, considering the criticisms about the rationality of static conjectural equilibria, it is to review the efforts made to connect the concept of conjectural variations with dynamic games, with or without complete information. Third, it is to propose, along the way, generalisations, extensions of classical results and suggest directions for future research. Since our focus is on theory, we shall not present the important strand of this literature mentioned in the previous paragraph, dealing with empirical and econometric investigations of conjectural variations.

The new interpretations of the CVE model

Some explanations are in order, as the CVE model has been the subject of much controversies among scholars. In the earliest descriptions, the conjectures were considered completely exogenous, a characteristic that very soon appeared to undermine the theoretical grounds of the CVE concept. Indeed, under the CVE concept the outcome of the game is configured by the conjectures, which means conversely that almost any observed behaviour can be described as a CVE with a suitable choice of conjectures. In other words, it has been realised that the CVE with arbitrary conjectures provided a theory of economic behaviours that was not refutable. Some refinements of the concept have then been proposed to overcome this Popperian criticism; it was conceded that the choice of conjectures should be treated as endogenous in the game. For instance, Laitner (1980), Bresnahan (1981), Perry (1982), Kamien and Schwartz (1983) and Boyer and Moreaux (1983a) have studied the idea of Consistent Conjectural Variations Equilibrium (CCVE). According to this concept, conjectures are required to be *consistent* in the sense that the best response functions obtained under those conjectures must correspond, to some extent, to the conjectured reaction functions.

Unfortunately, this line of argument is not sufficient to overcome another criticism, in relation with the foundations of solution concepts in game theory; as it will be explained in Chapter 1, in some static games, complete information and common knowledge of rationality rule out anything but Nash equilibria. This seriously reduces the set of situations where the concept of CVE could be consistently applied, and raises doubts about its generality.

To summarise, until recently it seems that the idea of conjectures in Game Theory had vanished either because it did not obey the discipline of any science to provide a refutable theory, or because its theoretical foundations appeared to be weak, at least in static situations of complete information and common knowledge.

Our use of the word 'seems' in the previous sentence suggests that one should be cautious before burying the idea of conjectural variations; actually we have witnessed in recent years a revival of this idea, for some new incarnations of it, using a dynamic point of view, are immune to all the previous objections. These new approaches are not designed to refute the objections addressed to the CVE concept; rather those objections point to difficulties that do not arise (or do not seem to arise) in the proposed settings.

First, several researchers have proposed to use the conjectural variations equilibrium as a *shortcut* for more complicated behaviours in implicit dynamic games (Dockner (1992); Cabral (1995); Itaya and Shimomura (2001); Itaya and Okamura (2003)). In this framework, agents have complete information and the assumption of common knowledge holds. But epistemic objections against CVE are irrelevant since only Nash equilibria are investigated. Static games have dynamic extensions; formal relationships are made explicit between the dynamic equilibrium concepts and the CVE of the associated static games.

Second, there is a growing literature based on the idea that agents, facing incomplete information, or having limited rationality, form conjectures (or beliefs) about the strategy of their opponents. In a repeated game situation, sufficiently rational agents will not only maximise their payoff given their beliefs, but will also check that their beliefs are consistent with observations. For instance Friedman and Mezzetti (2002), in the context of a dynamic oligopoly with boundedly rational firms, offer a logical reinterpretation of the conjectural variations model. Each firm solves a dynamic optimisation problem believing that the other firms will alter their future choices in proportion to its own current change, and these beliefs adapt in light of observed behaviours.

Organisation of the monograph

Our review of the theoretical literature is organised in four chapters.

The first chapter offers a description of the classical static conjectural variations equilibria. The concept of conjectural variations first appeared in static contexts of oligopoly (Bowley (1924); Frisch (1933)) and afterwards in the study of puzzling issues surrounding the private provision of public goods. One difficulty for the reader who discovers these articles stems from the variety of contexts and notations. In order to evacuate this unnecessary complication, Chapter 1 offers a survey of these static contributions within a unified mathematical notation. It first defines the main concepts: the conjectures, the conjectural variations, the reaction curves, and the corresponding *conjectural variations equilibria*. Then, we provide a classification and a comparison of the different definitions of conjectural variations equilibria that have been proposed in the literature. Results of existence are also reported. A particular attention is devoted to the requirement of *consistency*, as studied by Bresnahan (1981) for two players.

The chapter is complemented, at the end of the monograph, by appendices exposing the detailed treatment of mathematical and economic issues of CVE. Appendix A explains the geometric interpretation of CVE and exhibits general cases of existence of consistent conjectural equilibria. Appendix B offers qualitative results regarding the comparison of conjectural equilibria with Nash equilibria and the Pareto-efficient outcome, with a discussion of the welfare implications resulting from these comparisons. Appendix C discusses the examples of CVE appearing in the literature, in particular in the context of oligopoly theory, and in the theory of Public Goods.

The second chapter reviews some of the attempts at explaining how conjectural variations equilibria emerge as shortcuts from dynamic game models. The principle is that CVE capture dynamic behaviours (like threats of future punishments) within a static framework. Complicated interactions over time boil down to some kind of "reduced-form" static competition, formally equivalent to a static game with conjectural variations. Those kinds of equivalence have been investigated for the theory of contributions to a public good (Itaya and Shimomura (2001); Itaya and Okamura (2003)) and for the theory of the oligopoly (Dockner (1992); Cabral (1995)). In the same spirit, we offer a study of infrastructure competition between jurisdictions.

Another approach has been to formalise explicitly dynamic conjectures; the very idea of reaction beneath the concept of conjectural variations suggests to do so. Chapter 3 is devoted to the few papers that have proposed a definition of *consistent* equilibria for conjectures in a dynamic context. The definitions are reviewed and compared, whenever possible, since the concepts are used in different game situations. Fershtman and Kamien (1985) offer a new interpretation of open-loop and closed-loop equilibria in differential games as particular forms of CVE. Friedman (1977) and Laitner (1980) use a duopoly framework to extend to a dynamic context the ideas of consistent conjectures. In both cases a discrete-time infinite-horizon game problem is considered. Başar *et al.* (1986) propose a linear-quadratic, discrete-time dynamic game for the interaction of the monetary policies of two countries. Among several methods of solutions, feedback Nash equilibria and feedback Stackelberg behaviours, the authors propose to find a consistent conjectural variations equilibrium at each stage of the game. They call this equilibria "feedback consistent conjectural variations equilibria". In contrast with the approaches described above, the notion of consistency is weaker, being required at each period but not across time.

The ideas of global consistency of conjectures leading to technical difficulties, a last, and perhaps more promising approach, has been to use the ideas of conjectures for agents with a limited rationality and with incomplete information. Chapter 4 discusses several models, where the concept of conjectures is associated, within a dynamic model, to the idea of learning the "behaviour" of other agents. Itaya and Dasgupta (1995) show for a model of voluntary contributions to a public good, that a process with a dynamic improvement of the conjecture may converge in a locally stable fashion to the consistent conjectural variations equilibrium of the model. Friedman and Mezzetti (2002) introduce a dynamic, discrete-time, oligopoly model using adaptive beliefs. Players have the belief that their opponents will react in a linear fashion to *increments* of their strategies. They adapt these beliefs according to observed behaviours. One of the main results is that players move towards cooperation in steady state, when the discount parameter tends to one. Adapting the work of Friedman and Mezzetti, we propose a discrete-time learning model where players also have a linear conjecture about the behaviour of other players. Players optimise step-by-step their conjectured utility, and try to learn the proportionality factor of their opponents. It is shown that Pareto optimal strategies belong to the set of possible limits to this learning process.

Acknowledgements

Writing such a monograph was not our initial project when we discovered this literature. Our interest started with empirical papers where the authors have made use of the concept of conjectural variations to explain systematically observed departures from the prediction of the Nash equilibrium (Slade (1995), Bordignon (1994), Sugden (1982)). In a second step, trying to understand theoretical contributions on the subject, we ended up with a bulk of notes about many different papers, which we tried to organise within a survey. We are very grateful to Professor Ichiishi, editor of this series, for suggesting to write a monograph as an outgrowth of the seminal survey and for his patience (the task was significantly longer than we expected). We also thank Professors J.W. Friedman, J.J. Gabszewicz, G.J. Olsder and an anonymous reviewer of this manuscript for high-value criticisms and suggestions. We thank Professor R. Lifran for introducing us to conjectural variations, and his constant support in our work. We have also received a kind support from Professor E. Dockner in this enterprise. Finally, thanks are due to V. Fromion, C. Mezzetti and D. Claude for their careful reading of important parts of the monograph. Of course we keep responsibility for possible remaining flaws, and unavoidable, though usual, imperfections.

Contents

Chapter 1

Static Conjectural Variations Equilibria: Initial Concepts

1.1 Introduction

This chapter presents the different concepts of Conjectural Variations Equilibria (CVE) that have been proposed in static strategic settings. The main objective is to review the seminal results of the literature, to state them in a common and general notation and to document the existence of such equilibria (or non existence). Yet starting with such a general approach would have a number of drawbacks for the reader who discovers this literature, which we would like to avoid. Indeed, a general presentation is not the easiest way to get acquainted with the basic idea behind the concept of CVE. Nor it is the easiest way to point out different epistemic problems and flaws of the earlier developments of this concept. A way out is to start the exposition with a simple and carefully worked example, before turning to a general framework. We shall adopt this two-step approach in this chapter.

Section 1.2 makes use of the celebrated Cournot duopoly framework to present the initial concept of conjectural variations equilibrium, and to compare it with the Cournot-Nash equilibrium. This example is also used to broach the main developments of the concept, and to illustrate some epistemic issues. Then in Section 1.3 we go through formal definitions of the main concepts, with a particular attention to the Consistent Conjectural Variations Equilibrium (CCVE). We discuss the different variants of conjectural variations equilibria proposed over time in the literature, and their mathematical characterisations. Conjectural variations equilibria are then illustrated through several economic examples, in Section 1.4.

In Section 1.5, we suggest a formal framework generalising the ideas underlying conjectural variations equilibria, but allowing to formulate conjectures in a broader sense. Finally, Section 1.6 concludes and proposes future directions for investigation.

This chapter is far from being a comprehensive review of what can be said about static CVE. Many original complements to the topic will be found in Appendices A, B and C.

1.2 Origin of the conjectural variations concept

The concept of conjectural variations appeared in the context of Cournot's model of duopoly (Bowley (1924)). Two firms compete on the market of a homogeneous good. The decision variables are the quantities to be produced. Assume for simplicity that the inverse demand function is linear; if $q_1 \geq 0$ and $q_2 \geq 0$ denote the quantities produced by firms 1 and 2 respectively, then the market-clearing price p is:

$$p = a - (q_1 + q_2) ,$$

where a is a positive parameter. Admissible quantities are such that $p \geq 0$.

The two firms have the same technology. There are no fixed costs and the marginal cost is constant. Thus the cost to firm i to produce any given quantity is supposed to be a linear function:

$$C(q_i) = cq_i , \quad c > 0.$$

For firm i, the profit function is:

$$V^i(q_i, q_j) = [a - (q_i + q_j) - c] \, q_i , \quad j \neq i.$$

Quantities are chosen simultaneously, once and for all. Each firm behaves in order to maximise its profit function and, clearly, its decision problem is configured by the rival decision. Firm i's expectation about what the rival decision q_j could actually be is an information of paramount importance, for it drives its own choice q_i. Here appears the fundamental and distinctive assumption underlying the concept of conjectural variations equilibria. It says that for firm i the rival decision q_j is a function of its own choice q_i, formally $q_j = \mathbf{q}_j(q_i)$. This assumption deserves a careful discussion. We shall comment on it below, but for the time being let us investigate what prediction about optimal decisions can be made on this basis. Firm i's optimal output choice solves the following program:

$$\max_{q_i} V^i(q_i, \mathbf{q}_j(q_i)) .$$

An interior equilibrium satisfies the system of first order conditions

$$V_i^i(q_i, q_j) + V_j^i(q_i, q_j)\,\mathbf{q}_j'(q_i) \;=\; 0\ ,\ \ i = 1, 2,\ j = 1, 2,\ \ i \neq j\ ,$$

where V_k^i denotes the partial derivative derivative of V^i with respect to the variable q_k. We shall often use the notation f' for the derivative of the single-variable function f with respect to its variable, as in $\mathbf{q}_j'$.

More specifically in our example, the first order conditions are:

$$a - (2q_i + q_j) - c - q_i\mathbf{q}_j'(q_i) = 0\ ,\ \ i = 1, 2,\ j = 1, 2,\ \ i \neq j\ . \qquad (1.1)$$

The term $\mathbf{q}_j'(q_i)$ is referred to as the *conjectural variation* of firm i, *i.e.* firm i's conjecture about firm j's reaction to an infinitesimal change in q_i. It is worth noting that when this term is equal to zero (firm i expects the rival decisions to be independent of its owns) the first order conditions boil down to those of the Cournot-Nash equilibrium. Presumably, when this term is different from zero, conjectures give rise to a different outcome. This outcome is called the *Conjectural Variations Equilibrium* (or CVE) of the game.

More precisely in this example, any outcome ranging from the cartel solution to the perfect competition can arise with an adequate choice of the conjecture; the reader familiar with this model can check that the former occurs when $\mathbf{q}_j'(q_i) = 1$, whereas the latter follows when $\mathbf{q}_j'(q_i) = -1$.

From an epistemic point of view, this ability to predict anything is a serious flaw. The theory does not conform itself to the discipline of the falsificationist methodology, which is by now widely accepted among economists (see Hands (2001), Chapter 7). Blaug (1992), p. *xiii* explains that such a methodology "regards theories and hypotheses as scientific if and only if their predictions are at least in principle falsifiable, that is they forbid certain acts/states/events from occurring." This requirement is clearly violated when conjectures can be chosen arbitrarily. To overcome this early criticism, the concept was required to provide refutable predictions. A simple way out has been to consider endogenous conjectures. Economists first achieved that by imposing, roughly speaking, that the conjectural variations of each firm correspond to the slope of the *best response* of its opponent. The best response function of firm j, say, $\mathbf{b}_j(q_i)$, which results from j's mechanism of conjectures, is obtained as the solution of first order conditions (1.1), stated for firm j, with respect to its strategic variable q_j. This refinement of the concept was referred to as a *Consistent Conjectural Variations Equilibrium* (or CCVE).

Restricting attention to symmetric and *constant* conjectures, that is, assuming that $\mathbf{q}'_j(q_i) = r$ for any q_i and all i, it is found that in the above example, only the conjecture $r = -1$ is consistent. The fact that $\mathbf{q}'_j(q_i) = \mathbf{q}'_i(q_j) = -1$ is a consistent conjecture for both firms can easily be checked *ex post.* Substituting $\mathbf{q}'_i(q_j) = -1$ into the first order condition (1.1) for firm j, gives:

$$a - (2q_j + q_i) - c + q_j = 0 .$$

Accordingly, the best response of firm j, which is the solution of this equation in the variable q_j, is:

$$\mathbf{b}_j(q_i) = a - c - q_i .$$

The slope of the resulting best response is therefore $\mathbf{b}'_j = -1$, as conjectured. This CCVE predicts one and only one outcome here: the Bertrand outcome (or the perfect competition outcome).

Once the above Popperian criticism is discarded, this concept seems to be appealing on several counts. It explicitly captures the idea that actions are not independent of one another, and that oligopolists are concerned about one another's reactions. However, upon reflection it remains quite bizarre. It is intuitive that firm i's expectation about rival decisions should be based on the relevant pieces of information firm i possesses. Still, on what grounds should firm i believe the actual choice made by firm j depends on something firm j does not observe and about which it has no information prior making its choice? What does the concept of conjectural variations equilibrium means in terms of firm's information and rationality?

The more demanding assumptions, with regard to information and rationality one can make, are twofold. First there is *complete information*, meaning that each firm knows what are all the strategy spaces and all the profit functions (see for instance Gibbons (1992), Chapter 1). Second there is *common knowledge* of rationality: firm i knows that firm j is rational and is fully informed about strategies, its own and the rival profits, and the fact that the opponent is a rational decision maker as well, that firms i knows all that, and that firm i knows that firm j knows etc. A discussion of the common knowledge assumption can be found in Osborne and Rubinstein (1994), Chapter 5. On those grounds it makes sense that each firm can build a model for the competitor's behaviour and look for outcomes that are consistent with complete information and common knowledge.

A careful investigation reveals that, in this Cournot duopoly example, any outcome different from the Cournot-Nash equilibrium is inconsistent with the complete information and common knowledge assumptions. To show that, let us take the point of view of firm i and rather to ask what it should do, let us investigate what it should not do. First, it should not pick a strategy q_i where there exists another strategy q_i' giving a higher profit whatever the rival firm ever decides to choose: the rationality and complete information of firm i discard *dominated* strategies. Second, it can consider the rival's decision problem and understand whether some rival strategies will never be chosen because they are dominated (common knowledge). In the remaining space of strategies, some quantities may become dominated even though they were not initially. It is therefore possible to perform a second round of deletions of dominated strategies. Taking the argument a step further one may consider a third round of deletions, a fourth round, and so on until no remaining strategy is dominated. In other words an iterated process of deletion of dominated strategies can take place. Only those strategies that survive such a process are consistent with complete information and common knowledge. For our simple duopoly there is only one pair of strategies that survives this iterated process: the Cournot-Nash equilibrium.

This can be shown in a textbook manner (see Gibbons (1992), Chapter 1 or Fudenberg and Tirole (1991), Chapter 2 for further details). The first point is to observe that, in our context, a quantity q_i is dominated if and only if there is no choice q_j of firm j for which the strategy q_i is a best response. Then remark that any strategy higher than a is dominated by the strategy to produce 0. Indeed the price, and therefore profits, are negative if a firm chooses a quantity higher than a. Third, considering firm i's best response, see that the best response to $q_j = 0$ is $q_i = (a-c)/2$, the monopoly quantity, and then the best responses decline as q_j increases: thus for any $q_j \geq 0$, firm i's best responses are less or equal to $(a-c)/2$ (see Figure 1.1, where the best response function of firm i is denoted as $BR_i(q_j)$). Finally, given this upper bound on firm i's quantities it is possible to deduce a lower bound on firm j's quantities. Indeed the best response for firm j to firm i's upper bound strategy $(a-c)/2$ is $(a-c)/4$, and then this best response increases as the q_i decreases from $(a-c)/2$. Then given this lower bound it is possible to find a new upper bound for firm i's quantity and so on... continue this way to end up in the (Cournot) Nash equilibrium.

The conclusion then follows: for a conjectural variations equilibrium of any kind (consistent or not, but with non-zero conjectures) to make sense in this static framework it is necessary that either the firms don't have

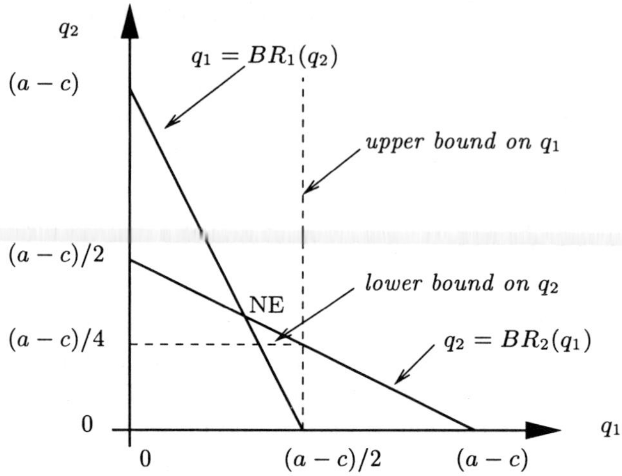

Fig. 1.1 Best Responses in a symmetric duopoly

complete information, or the common knowledge assumption is relaxed, or both possibilities. A third possibility would be that the static model itself does not properly render the game situation at hand, and that a fully dynamic formulation would be preferable. The conjectures are then an attempt to incorporate true reactions in a static setting.

The above argument does not mean that a CVE is a nonsense in any conceivable static game of complete information and common knowledge. But, it definitely means that the set of static situations where it can be used consistently is reduced, at least as compared with the Nash equilibrium. To summarise, the conjectures that could lead to equilibria made of dominated strategies (directly or indirectly) are in conflict with the assumptions of complete information and common knowledge. For other situations, there are no arguments against CVE of which we are aware. Readers interested about the links between various assumptions concerning information, knowledge and solution concepts can find a non-technical introduction in Board (2002).

The view developed in the present monograph is that there is a renewal of interest for the concept of conjectures in game situations, and this renewal is in relation with the absence of complete information, common knowledge, or proper dynamic formulations. Static conjectural variations models, it is argued, can be useful *shortcuts* to capture in a simple way the messages of more complicated, but properly specified models. There are therefore good reasons for studying static conjectural variations models as we do in this chapter, as shorthands, provided that the reader is aware of the existence of limitations and dangers with such an exercise.

1.3 Definitions and characterisation of Conjectural Variations Equilibria

This section presents static games. It gives definitions for the equilibria that can be considered for these games when taking into account the possible reactions of the other players: conjectural variations equilibria (Definition 1.3) and consistent conjectural variations equilibria (Definition 1.4). The different variants existing in the literature for each concept are surveyed, and we discuss existence properties for each of them.

1.3.1 *Notation and assumptions*

The remainder of this chapter considers game-theoretic situations in which n players try to maximise their payoff function. The set of admissible strategies e_i for player i will be denoted by E_i, and the set of admissible profiles of strategies $e = (e_1, \dots, e_n)$ will be $\mathbb{E} = E_1 \times \dots \times E_n$. The E_i are assumed to be open sets, since we will be interested only in interior solutions of the optimisation problems.[1] Finally, the payoff function of player i will be $V^i\colon \mathbb{E} \to \mathbb{R}$.

Conjectural variations are defined in a differentiable context. We shall therefore assume that E_i is a connected subset of the real line,[2] and that payoff functions are regular enough.

For a real-valued function f defined on the strategy space, the notation $f_i = \partial f / \partial e_i$ stands for the partial derivative with respect to the variable e_i. We shall generally adopt a standard convention of the literature, denoting $e_{-i} = (e_j; j \neq i)$ the profile of the strategies of player i's opponents, and placing the strategy of player i as the first argument of functions related to player i, as in: $V^i(e_i, e_{-i})$.

[1] Games resulting from the modelling of an economic situation often have a natural strategy space which is a closed set. Such are the illustrations used throughout this monograph. The results of this chapter can be applied using the topological *interior* of those strategy spaces, which are an open sets. The theory of conjectural variations has not been extended to "corner solutions" so far.

[2] The current literature on conjectural variations equilibria seems to restrict itself to single-dimensional strategy spaces. Apparently, a generalisation to many-dimensional spaces has not been attempted yet, but should be straightforward.

1.3.2 *Nash equilibrium, Pareto optimality*

We begin by recalling the classical definitions of Nash equilibria and Pareto optima, which shall be used throughout the monograph, for the purpose of future comparisons with these particular outcomes.

Definition 1.1 (NASH EQUILIBRIUM) A Nash equilibrium is a strategy profile $(e_1^N, \ldots, e_n^N) \in \mathbb{E}$ such that:

$$V^i(e_i^N, e_{-i}^N) \geq V^i(e_i, e_{-i}^N) , \qquad \forall e_i \in E_i, \ \forall i .$$

Definition 1.2 (PARETO OPTIMUM) A strategy profile $(e_1^P, \ldots, e_n^P)$ is a Pareto optimum if there does not exist another strategy profile $e = (e_1, \ldots, e_n) \in \mathbb{E}$ such that simultaneously: $V^i(e_1, \ldots, e_n) \geq V^i(e_1^P, \ldots, e_n^P)$ for every player i, with a strict inequality for at least one player.

1.3.3 *Conjectures, reactions and consistency*

The literature on conjectural variations has focused mainly on two-player games. Unless explicitly mentioned, so does this chapter, where i and j are used as the identities of the players, with the convention that $i \neq j$ if both appear in the same formula. A noticeable exception is Section 1.3.9 which is devoted to games with any number of players.

The central concept in the theory is the notion of *conjecture*. The different definitions of conjectural variations equilibria basically differ in the way the players form their conjectures. Generally, player i's conjecture is defined by means of a differential equation. The central concept is the *variational conjecture* r_j, which describes player j's reaction, as anticipated by player i, to an infinitesimal *variation* of player i's strategy. This mechanism leads to the notion of a *conjectured reaction function* of the opponent.

Given this conjectured reaction on the part of the opponent, each player optimises her perceived payoff. This leads to the concept of a *conjectural best response* function. The conjunction of the optimisation processes of both players results in some strategy profile. An equilibrium is obtained when no player has an interest in deviating from her strategy. The strategy of each player is then the conjectural best response to the strategy of the other player.

The *consistency* (or, sometimes, "rationality") of the equilibrium is defined as the coincidence between the *conjectural best response* of one player with the *conjectured reaction function* for that other player. This coincidence can be defined using several degrees of strength. The weakest definition of consistency that has been proposed in the literature requires that the coincidence holds only at the equilibrium. A stronger definition requires that the coincidence holds in a neighbourhood of the equilibrium. We therefore distinguish these "punctually" consistent conjectures from the more strongly consistent conjectures.

In the literature, and for two-player games, the conjectural variations take two forms:

i) player i considers that the variation of player j's strategy, r_j, depends on strategies of all players: $r_j(e_i, e_j)$; outcomes in this case will be referred to as: General Conjectural Variations Equilibria (GCVE).

ii) player i considers that the variation of player j's strategy, depends only on her own strategy and has the form: $r_j(e_i)$; the corresponding outcomes will be called: Conjectural Variations Equilibria (CVE).

Which of these forms should the modeller select in practice? Clearly, the most general form $r_j(e_i, e_j)$ should be preferred *a priori*. As argued by Boyer and Moreaux (1983b), p. 29, in the absence of further information on the actual behaviour of her opponent, a player trying to summarise this behaviour in a conjectural variation should take no chances and adopt a model with as many parameters as practically feasible. Following this line of thought, if the specific functional form of the conjectural variations is to be determined by observations, econometric tests should decide whether parameters associated to one or the other player should be taken as zero, or not.

On the other hand, papers studying conjectures of the form $r_j(e_i)$ seem to adopt the idea that each player somehow thinks she has a position of leader (*à la* Stackelberg) and that her opponent does *react* to her play. The modelling of the opponent as a function $r_j(e_i)$ is natural in this context.

From a mathematical standpoint, using one-variable conjectures obviously simplifies the analysis. When the issue of consistency is addressed, the literature mostly restricts its attention to constant conjectures, with no variables at all.

The different concepts and the corresponding acronyms are summarised in the following table:

Form of variational conjectures	$r_j(e_i, e_j)$	$r_j(e_i)$
Definition of equilibria	GCVE (§ 1.3.4)	CVE (§ 1.3.5)
Consistency	CGCVE (§ 1.3.6)	CCVE (§ 1.3.7)
Consistency at the equilibrium	LRCE (§ 1.3.8)	

We now proceed to give precise definitions for these concepts, provide characterisations of equilibria and consistent equilibria in the form of systems of (functional) equations, and discuss existence results.

1.3.4 *Conjectural Variations Equilibria with general conjectures (GCVE)*

The less restrictive definition for conjectural variations equilibria is the one used by Laitner (1980), Ulph (1983), Boyer and Moreaux (1983a), Boyer and Moreaux (1983b) (in the context of duopoly theory), or Cornes and Sandler (1984) (in the context of the theory of public goods).

1.3.4.1 *Definitions*

Conjectural variations equilibria are based on the idea that players consider possible variations of the strategy of their opponent. Where there is a variation, there must be some "initial" strategy from which the variation is contemplated. This leads to the notion of a *benchmark* (or "reference") strategy profile.

In relation to any such benchmark profile $e^b = (e_1^b, e_2^b) \in \mathbb{E}$, player i conjectures that an infinitesimal variation δe_i of her strategy will be followed by a variation of player j's strategy $\delta e_j = r_j(e_i^b, e_j^b)\delta e_i$. Considering now arbitrary (non-infinitesimal) variations, player i is led to think that if she plays strategy e_i, player j's strategy is: $e_j = \rho_j^c(e_i; e_i^b, e_j^b)$, where the function ρ_j^c is the (assumed unique) solution of the ordinary differential equation in the variable e_i:

$$\frac{\partial \rho_j^c(e_i; e_i^b, e_j^b)}{\partial e_i} = r_j(e_i, \rho_j^c(e_i; e_i^b, e_j^b)) , \tag{1.2}$$

with initial condition $\rho_j^c(e_i^b; e_i^b, e_j^b) = e_j^b$. Indeed, if player i does not deviate at all from her assumed benchmark strategy e_i^b, she assumes that player j will not deviate either, and play e_j^b.

This function $\rho_j^c(e_i; e_i^b, e_j^b)$ will be called the *conjectured reaction function* for the conjectural variation $r_j(e_i, e_j)$, for a given benchmark strategy profile (e_i^b, e_j^b).

By assumption, each player optimises her payoff. Given the benchmark strategy (e_i^b, e_j^b) and the conjectured reaction function ρ_j^c, player i should solve the following optimisation problem:

$$\max\{V^i(e_i, e_j) \mid (e_1, e_2) \in \mathbb{E} \text{ and } e_j = \rho_j^c(e_i; e_i^b, e_j^b)\} . \tag{1.3}$$

For each benchmark strategy profile, this process yields a strategy for player i which is perceived as optimal. Which strategy profile should the players choose as benchmark? This profile should be so that no player thinks she has an interest to deviate from it, still given the conjecture made on the opponent. In other words, the benchmark strategy profile should itself be the solution of the maximisation problem (1.3). We have therefore the definition:

Definition 1.3 (General conjectural variations equilibrium) A pair of variational conjectures $r_i(e_j, e_i)$ $i = 1, 2$, together with a pair of strategies $(e_1^c, e_2^c) \in \mathbb{E}$ is a General Conjectural Variations Equilibrium (GCVE) if (e_1^c, e_2^c) is a solution of the optimisation problem:

$$\max\{V^i(e_i, e_j) \mid (e_1, e_2) \in \mathbb{E} \text{ and } e_j = \rho_j^c(e_i; e_i^c, e_j^c)\} , \tag{1.4}$$

simultaneously for $i = 1, 2$.

This optimisation problem is reminiscent of what happens in a Stackelberg game. In that situation the leader of the game, being the first to play, optimises her payoff taking into account the rivals reaction, which she can deduce. Indeed, she knows that the rival (the follower) is rational and will optimise her payoff. The leader can therefore replace, in her payoff function, the rival's variable with her *Nash best response* function χ_j^N. This leads to an optimisation problem formally similar to (1.4), with ρ_j^c replaced by χ_j^N.

In a conjectural variations game, *both* players can be seen as acting as leaders. This explains why conjectural variations are sometimes described as a "double Stackelberg" game. An important difference is that a Stackelberg leader derives her conjecture from her information about follower; in a CVE the treatment of players' information is unclear. The following chapters of this monograph will discuss how this apparently irrational behaviour can emerge in a dynamic setting.

1.3.4.2 *Characterisation of GCVE*

Definition 1.3 is not directly appropriate for computing equilibria. The following theorem provides the proper tools.

Theorem 1.1 *Assume that V^i is twice differentiable.*

i) If (r_1, r_2) and (e_1^c, e_2^c) is a GCVE, then it satisfies

$$e_i^c = \chi_i(e_j^c) \qquad i \neq j \,, \tag{1.5}$$

where the function $\chi_i(e_j)$ is implicitly defined by the solution of the following first order conditions for each player:

$$V_i^i(e_i, e_j) \; + \; r_j(e_i, e_j)\, V_j^i(e_i, e_j) = 0 \,. \tag{1.6}$$

ii) Conversely, for given conjectures (r_1, r_2), any pair (e_1^c, e_2^c) solution of (1.5)–(1.6) is a GCVE if the following maximality condition holds:

$$\frac{d^2\overline{V}^i}{d^2 e_i}(e_i^c) < 0 \,, \tag{1.7}$$

for each player, where $\overline{V}^i$ is defined by

$$\overline{V}^i(e_i) \;=\; V^i\left(e_i, \rho_j^c(e_i; e_i^c, e_j^c)\right) \,. \tag{1.8}$$

Proof. Player i has to solve Problem (1.4). Under the constraint on e_j, the function of a single variable (1.8) is to be maximised. Looking for the first order condition, and differentiating with respect to e_i, we obtain:

$$\begin{aligned}\frac{\mathrm{d}\overline{V}^i}{\mathrm{d}e_i}(e_i) &= V_i^i\left(e_i, \rho_j^c(e_i; e_i^c, e_j^c)\right) \; + \; r_j(e_i, \rho_j^c(e_i; e_i^c, e_j^c))\; V_j^i\left(e_i, \rho_j^c(e_i; e_i^c, e_j^c)\right)\\ &= 0 \,.\end{aligned}$$

Condition (1.7) corresponds to the usual sufficient condition for maximisation. Hence the conclusion. □

It is usual in the literature to find Theorem 1.1 stated as the definition of GCVE. This point of view tends to eclipse the fact that there is an optimisation process at work.

The function $e_i = \chi_i(e_j)$ defined as the solution of Equation (1.6) is called the *conjectural best response.* Observe that the introduction of this function is not essential to the definition of the equilibrium. Indeed, given that $e_i^c = \chi_i(e_j^c)$, $i, j = 1, 2$, the problem amounts to solving a system of two equations with two unknowns (1.6). On the other hand, it is interesting to identify these functions, since they are analogous to the usual "best response functions" in economic theory. Indeed, consider the situation of the classical "Nash play". Nash equilibria are particular cases of GCVE, where the conjectural variations each player anticipates on the part of her opponent are identically zero: $r_j \equiv 0$. Then Equation (1.6) boils down to $V_i^i(e_i, e_j) = 0$, and the solution $e_i = \chi_i^N(e_j)$ of this equation is the Nash best response of player i. As we shall see, conjectural best response functions are essential for the definition of consistency in Section 1.3.6.

Conjectured reaction functions and conjectural best response functions are not to be confused. All these functions have the same value at the equilibrium, since, by definition, $\rho_i^c(e_j^c; e^c) = \chi_i(e_j^c)$. Geometrically speaking, these four curves (two for each player) pass through the equilibrium strategy profile. We come back to the geometric properties of the equilibria in Appendix A.

Another point to stress is the importance of the benchmark strategy profile e^b in the definition. At the equilibrium, each player considers (conjectural) variations with respect to the equilibrium itself, so that there is no necessity to imagine possible deviations with respect to other strategy profiles. Indeed, as we have just observed, it is possible to define GCVE without the device of general benchmark strategy profiles, using directly the first order conditions of the equilibrium. However, *out of equilibrium*, the question of the benchmark strategy is essential. We come back to this point when we study dynamic game models (Chapter 3) and learning models (Chapter 4).

1.3.4.3 *Existence results*

By choosing adequately the functions r_i, it is possible to obtain any strategy $(\bar{e}_1, \bar{e}_2) \in \mathbb{E}$ as a GCVE solution of (1.6), at least if $V_j^i(\bar{e}_i, \bar{e}_j) \neq 0$. Indeed, if the function r_j is such that, at the particular strategy profile:

$$r_j(\bar{e}_i, \bar{e}_j) \;=\; -\,\frac{V_i^i(\bar{e}_i, \bar{e}_j)}{V_j^i(\bar{e}_i, \bar{e}_j)}\;,$$

then it satisfies the first order condition (1.6). It is possible to construct functions r_j such that the second order condition (1.7) holds in the neighbourhood of $(\bar{e}_1, \bar{e}_2)$.

In particular, Pareto optima can be conjectural equilibria if they satisfy the above condition.

1.3.5 *Conjectural Variations Equilibria (CVE)*

When the variational conjecture of player i depends only on her own strategy e_i, one obtains what is simply called "conjectural variations equilibria". This notion is used by Bresnahan (1981); Bresnahan (1983); Perry (1982); Itaya and Dasgupta (1995) or Sterdyniak and Villa (1993).

Since r_j depends only on e_i, the conjectured reaction functions, solution of (1.2), are of the form:

$$\rho_j^c(e_i; e_i^b, e_j^b) \;=\; e_j^b \;+\; \int_{e_i^b}^{e_i} r_j(u)\,\mathrm{d}u\;. \tag{1.9}$$

As already noticed by Robson (1983) (see also Laitner (1980)), considering the variational conjecture r_j is equivalent to considering as conjecture the *whole family* of conjectured reaction functions ρ_j^c.

Substituting $r_j(e_i)$ for $r_j(e_i, e_j)$ and using the form (1.9) for ρ_j^c in Definition 1.3 and Theorem 1.1, we obtain the definition of a Conjectural Variations Equilibrium (CVE), and the corresponding characterisation. In particular, if a pair of variational conjectures (r_1, r_2) and a strategy profile (e_1^c, e_2^c) form a CVE, then (e_1^c, e_2^c) solves the first order condition

$$V_i^i(e_i, e_j) \;+\; r_j(e_i)V_j^i(e_i, e_j) \;=\; 0\;, \tag{1.10}$$

simultaneously for $i = 1, 2$.

Concerning existence results, the argument of Paragraph 1.3.4.3 still holds: provided that $V_j^i(\bar{e}_i, \bar{e}_j) \neq 0$, any strategy $(\bar{e}_1, \bar{e}_2) \in \mathbb{E}$ can be a CVE. It is sufficient to choose $r_j(e_i)$ continuous such that:

$$r_j(\bar{e}_i) \;=\; -\;\frac{V_i^i(\bar{e}_i, \bar{e}_j)}{V_j^i(\bar{e}_i, \bar{e}_j)}\;.$$

This extreme multiplicity of equilibria, resulting from the exogeneity of the conjecture function, has prompted authors to devise a mechanism by which conjectures would result from an endogenous reasoning. This has led to the concept of consistency which we develop in the next sections.

1.3.6 *Consistent General Conjectural Variations Equilibria (CGCVE)*

1.3.6.1 *Definition*

In terms of conjectural variations, we have seen that consistency amounts to requiring that (conjectural) best response functions be equal to conjectured reaction functions. This amounts in turn to saying that the conjectural best response function is a solution of the differential equation (1.2). We have therefore the definition (see Laitner (1980); Ulph (1983)):

Definition 1.4 (CONSISTENT GENERAL CONJECTURAL VARIATIONS EQUILIBRIUM) A pair of strategies (e_1^c, e_2^c) and the variational conjectures $r_i(e_1, e_2)$, $i = 1, 2$ are a Consistent General Conjectural Variations Equilibrium (CGCVE) if

i) (e_1^c, e_2^c) is a GCVE for the variational conjectures (r_1, r_2);

ii) $\chi_i(e_j)$ being a solution in e_i of Equation (1.6), then for some $\varepsilon > 0$, and for $i = 1, 2$,

$$\chi_i'(e_j) = r_i(\chi_i(e_j), e_j) \ , \qquad |e_j - e_j^c| < \varepsilon \ .$$

Observe that this definition requires the coincidence of slopes in a neighbourhood of the equilibrium. This is in accordance with the seminal idea that only small variations about some reference point are relevant.

Alternately however, the coincidence could be required over the whole strategy space, as in Olsder (1981). The question of whether this is actually a strictly stronger requirement does not seem to have been investigated. In the case where conjectured reaction functions are linear (constant variational conjectures) and if conjectural best response functions are linear as well, the existence problem is the same with the local and the global formulations. This case is frequently encountered in the current literature.

1.3.6.2 *Characterisation of CGCVE*

Given Definition 1.4 and Theorem 1.1, we have the following characterisation. Since the focus is on the equilibrium strategy profile e^c, we simplify the notation and use $\rho_i^c(e_j)$ as a shorthand for $\rho_i^c(e_j; e^c)$ in the remainder of this section.

Theorem 1.2 *A pair of strategies (e_1^c, e_2^c) and the variational conjectures $r_i(e_1, e_2)$, $i = 1, 2$ are a Consistent General Conjectural Variations*

Equilibrium if and only if the conjectured reaction functions $(\rho_1^c(e_2), \rho_2^c(e_1))$ (solutions of (1.2)), satisfy:

$$e_1^c = \rho_1^c(e_2^c) , \quad e_2^c = \rho_2^c(e_1^c) ,$$

and there exists $\varepsilon > 0$, such that for $i, j = 1, 2$:

$$V_i^i(e_i, e_j) + r_j(e_i, e_j) \, V_j^i(e_i, e_j)\big|_{e_i=\rho_i^c(e_j)} = 0 \qquad \forall e_j, \quad |e_j - e_j^c| < \varepsilon . \tag{1.11}$$

Proof. Assume that (e_1^c, e_2^c) and $r_i(e_1, e_2)$ form a CGCVE. Then, by Definition 1.4, both functions $\chi_i(e_j)$ and $\rho_i^c(e_j)$ are solution of the differential equation (1.2). Since we have assumed that the solution is unique, we have:

$$\chi_i(e_j) = \rho_i^c(e_j) ,$$

in an ε-neighbourhood of e_j^c. Hence, in this neighbourhood, the function $\rho_i^c(e_j)$ is solution of (1.6). Therefore, (1.11) holds.

Conversely, if $\rho_i^c(e_j)$ is a solution of (1.11), then (assuming the uniqueness of the solution) it coincides with $\chi_i(e_j)$, solution of (1.6). Therefore, $\chi_i(e_j)$ is solution of the differential equation (1.2), which amounts to saying that:

$$\chi_i'(e_j) = r_i(\chi_i(e_j), e_j) ,$$

in a neighbourhood of e_j^c. This means that (e_1^c, e_2^c) and (r_1, r_2) are a CGCVE. □

1.3.6.3 *Existence results*

In general, there can be many consistent equilibria with general conjectures. Laitner (1980), Proposition II constructs (in the case of a duopoly) variational conjectures (r_1, r_2) such that any pair (e_1^c, e_2^c) (satisfying a reasonable condition) is a CCVE for these conjectures.

1.3.7 *Consistent Conjectural Variations Equilibria (CCVE)*

When the conjecture of player i about player j, r_j, depends only on e_i (her own strategy), we obtain a more restrictive concept of consistent equilibrium.

1.3.7.1 *Definition*

This definition is due to Bresnahan (1981). It is a particular case of Definition 1.4, but it is worth recalling it here since it is often encountered in the literature. According to Kamien and Schwartz (1983), the method for calculating the CCVE is originally due to Holt (1980). The mention of this concept appears independently in van der Weel (1975) and Olsder (1981).

Definition 1.5 (CONSISTENT CONJECTURAL VARIATIONS EQUILIBRIUM) A Consistent Conjectural Variations Equilibrium (CCVE) is a pair of strategies (e_1^c, e_2^c) and conjectures $(r_1(e_2), r_2(e_1))$ such that:

i) (e_1^c, e_2^c) is a CVE for the variational conjectures $(r_1(e_2), r_2(e_1))$ (Section 1.3.5);

ii) if $\chi_i(e_j)$, $i = 1, 2$, is the solution of (1.10) in e_i, there exists $\varepsilon > 0$, such that

$$\begin{aligned} r_2(e_1) &= \chi_2'(e_1) \qquad \forall e_1, \quad |e_1^c - e_1| < \varepsilon \\ r_1(e_2) &= \chi_1'(e_2) \qquad \forall e_2, \quad |e_2^c - e_2| < \varepsilon. \end{aligned}$$

1.3.7.2 *Characterisation of CCVE*

Theorem 1.2 applied to the particular conjectures under consideration yields a characterisation involving solely conjectured reaction functions (see Başar and Olsder (1999)):

Theorem 1.3 *A pair of strategies (e_1^c, e_2^c) and the variational conjectures $(r_1(e_2), r_2(e_1))$ are a Consistent Conjectural Variations Equilibrium if the conjectured reaction functions $(\rho_1^c(e_2), \rho_2^c(e_1))$ (solutions of the differential equation (1.9)), satisfy:*

$$e_1^c \;=\; \rho_1^c(e_2^c), \quad e_2^c \;=\; \rho_2^c(e_1^c)\ ,$$

and there exists $\varepsilon > 0$, such that for $i, j = 1, 2$:

$$V_i^i(e_i, e_j) + (\rho_j^c)'(e_i)\ V_j^i(e_i, e_j)\Big|_{e_i=\rho_i^c(e_j)} \;=\; 0 \qquad \forall e_j, \quad |e_j^c - e_j| < \varepsilon\ . \tag{1.12}$$

Differentiating Equation (1.12) with respect to e_j, additional sufficient conditions can be derived:

Corollary 1.4 *A CCVE with strategies (e_1^c, e_2^c) and reaction functions $(\rho_1^c(e_2), \rho_2^c(e_1))$ satisfies, for $i = 1, 2$:*

$$0 = \left(1 + (\rho_i^c)'(e_j)(\rho_j^c)'(e_i)\right) V_{ij}^i(e_i, e_j) \; + \; (\rho_i^c)'(e_j) V_{ii}^i(e_i, e_j) \\ + \; (\rho_j^c)'(e_i) V_{jj}^i(e_i, e_j) \; + \; (\rho_i^c)'(e_j) \, (\rho_j^c)''(e_i) \, V_j^i(e_i, e_j) \; , \quad (1.13)$$

with $e_i = \rho_i^c(e_j)$, for any e_j such that $|e_j^c - e_j| < \varepsilon$. In particular, for conjectures with constant variations $(\rho_i^c)'(\cdot) = r_i$ and $(\rho_j^c)'(\cdot) = r_j$, we have necessarily:

$$(1 + r_i r_j) \; V_{ij}^i(e_i, e_j) + r_i \; V_{ii}^i(e_i, e_j) + r_j \; V_{jj}^i(e_i, e_j) \; = \; 0 \; , \qquad (1.14)$$

for any e_j such that $|e_j^c - e_j| < \varepsilon$, and $e_i = \rho_i^c(e_j)$.

This has in turn an interesting qualitative property on the value and the sign of consistent constant variations conjectures:

Corollary 1.5 *Assume a symmetric game with identical, constant, consistent, conjectural variations, $r_1 = r_2 = r$, and equilibrium (e^c, e^c). Then, r has zero, one or two possible values. In the latter case, the values r_+ and r_-, are such that:*

$$r_+ r_- = 1 \qquad \text{and} \qquad \mathit{sign}(r_+) = \mathit{sign}(r_-) = \mathit{sign}\left(-\frac{V_{ii}^i + V_{jj}^i}{V_{ij}^i}(e^c) \right) \; .$$

Proof. According to (1.14), we have in particular that r is solution of the quadratic polynomial:

$$V_{ij}^i(e^c) \; + \; r \; (V_{ii}^i(e^c) + V_{jj}^i(e^c)) \; + \; r^2 \; V_{ij}^i(e^c) \; = \; 0 \; .$$

If the roots r_+ and r_- are real, their product is 1 and they have therefore the same sign, which is the sign of their sum. Hence the conclusion. □

1.3.7.3 *Existence results*

Existence results for CCVE in the literature are of two sorts: positive, and negative.

Positive results exist for particular models. In various papers, authors prove the existence (and uniqueness among economically feasible strategies) of a consistent *constant* CVE. Some of the economic models where CCVE are known to exist are reviewed in Sections 1.4 and Appendix C. In Section

A.2 of the Appendix, we exhibit general families of payoff functions that can have consistent conjectural equilibria.

Negative results are also partial: for the symmetric duopoly situation (the model discussed in Section 1.2; see also Section 1.4.1) but quadratic marginal costs, Bresnahan (1981) proves that there are no consistent polynomial conjectures. We discuss in Section A.3 the generalisation of Bresnahan's argument about the construction of polynomial conjectures. Robson (1983) extends the result of Bresnahan (1981), by showing that there are no (symmetric) analytic conjectures, in the same model. These negative results on one of the simplest economic models have cast some doubts on the relevance of the concept of CCVE. Considering the general existence results obtained by Olsder (1981), we believe however that the case is not closed, and that it would be interesting to revisit the problem.

1.3.8 *Equilibria with punctual consistency*

Boyer and Moreaux (1983a); Boyer and Moreaux (1983b) introduced a concept called *locally rational conjectural equilibria (LRCE)*.[3] It consists in requiring that the respective slopes of the conjectured reaction function and the conjectural best response function coincide only *at the equilibrium*. See also the "partial consistency" of Example 5 of Bresnahan (1981), for conjectures with one variable. This notion of consistency is therefore weaker than that of CGCVE.

1.3.8.1 *Definition*

For general payoff functions the definition is the following.

Definition 1.6 (LOCALLY RATIONAL CONJECTURAL VARIATIONS EQUILIBRIUM) A pair (e_1^c, e_2^c) and the variational conjectures $r_i(e_1, e_2)$, $i = 1, 2$ are a LRCE if

i) (e_1^c, e_2^c) is a GCVE for the variational conjectures (r_1, r_2);
ii) if we call $\chi_i(e_j)$ the solution of Equation (1.6) in e_i then:

$$\chi_i'(e_j^c) = r_i(e_i^c, e_j^c) \quad i = 1, 2, \quad i \neq j. \tag{1.15}$$

[3]In Mathematics, the vocable "locally" usually means "in a neighbourhood". This term should rather be attached to the concept of Definition 1.4, and the equilibrium discussed here should be named "punctually" consistent. We shall however conform to the current terminology.

In Başar and Olsder (1999) one finds the definition of a family of what can be called *punctually consistent* equilibria. This can be seen as a weakening of the notion of CCVE, based on the following local analysis around the candidate equilibrium (e_1^c, e_2^c):

$$\rho_i^c(e_j) = e_i^c + \frac{\mathrm{d}\rho_i^c}{\mathrm{d}e_j}(e_j^c)(e_j - e_j^c) + \frac{1}{2}\,\frac{\mathrm{d}^2\rho_i^c}{\mathrm{d}e_j^2}(e_j^c)(e_j - e_j^c)^2 + \frac{1}{6}\,\frac{\mathrm{d}^3\rho_i^c}{\mathrm{d}e_j^3}(e_j^c)(e_j - e_j^c)^3 + \ldots$$

Now, using the above expression for $\rho_i^c(e_j)$ around $e_j = e_j^c$ one finds new expressions, replacing Equation (1.12) as consistency requirement. For the zeroth order

$$V_i^i(e_i^c, e_j^c) + (\rho_j^c)'(e_i^c)V_j^i(e_i^c, e_j^c) = 0 , \tag{1.16}$$

and for the first order (arguments at $e_1 = e_1^c, e_2 = e_2^c$):

$$V_{ii}^i\,\frac{\mathrm{d}\rho_i^c}{\mathrm{d}e_j} + V_{ij}^i\,(1 + \frac{\mathrm{d}\rho_i^c}{\mathrm{d}e_j}\frac{\mathrm{d}\rho_j^c}{\mathrm{d}e_i}) + V_{jj}^i\,\frac{\mathrm{d}\rho_j^c}{\mathrm{d}e_i} + V_j^i\,\frac{\mathrm{d}^2\rho_j^c}{\mathrm{d}e_i^2}\frac{\mathrm{d}\rho_i^c}{\mathrm{d}e_j} = 0 . \tag{1.17}$$

Equation (1.17) is to be compared with (1.13). The difference is now that the equality is not required in the neighbourhood of the equilibrium, but only at the equilibrium. This leads to the following definition.

Definition 1.7 (CONSISTENT CONJECTURAL VARIATIONS EQUILIBRIUM OF ORDER 0 AND 1) A pair of strategies (e_1^c, e_2^c) and of conjectural variations (r_1, r_2) form a CCVE of order zero if the corresponding conjectured reaction functions (ρ_1^c, ρ_2^c) satisfy (1.16). It is called a CCVE to the first order if (1.16)–(1.17) are satisfied.

The definition can be generalised to any finite order, pursuing the Taylor expansion of Equation (1.12), and provided that the V^i are continuously differentiable up to that order. See Başar and Olsder (1999) for details.

With this definition, CCVE "up to some order" are a relaxation of CCVE in the following way: an equilibrium of order zero corresponds to a CVE (see Theorem 1.1 adapted to single-variable conjectures). Provided that the Taylor series converge in a neighbourhood of the strategy e^c, an equilibrium of infinite order would be a CCVE (see Theorem 1.3).

In summary, we have the following inclusions between sets of equilibria:

$$\begin{array}{ccccccc}
\textbf{CCVE} & \subset & \textbf{CGCVE} & \subset & \textbf{LRCE} & \subset & \textbf{GCVE} \\
\cap & & & & & & \cup \\
\textbf{CCVE order } p+1 & \subset & \textbf{CCVE order } p & \subset & \textbf{CCVE order } 1 & \subset & \textbf{CVE}
\end{array}$$

1.3.8.2 *Existence results*

The definition of LRCE is so weak that for the case of the duopoly, provided some technical conditions hold, there is an infinity of LRCE (see Boyer and Moreaux (1983b), Proposition 2).

No results concerning CCVE of order $p > 0$ are known.

1.3.9 *Conjectures in many-player games*

What about games with several players? There are no technical difficulties for establishing the notion of Conjectural Variations Equilibria for more than two players. Going back to Definition 1.3, one assumes that each player i thinks that *every other player* j will react to a small variation δe_i of her strategy by a variation $r_{ij}\delta e_i$, all variations being recorded with respect to a benchmark strategy profile e^b.[4] This defines the conjectured reaction functions ρ^c_{ij} by a differential equation similar to (1.2):

$$\frac{\partial \rho^c_{ij}(e_i; e^b)}{\partial e_i} = r_{ij}(e_i, \rho^c_{-i}(e_i; e^b)) ,$$

where ρ^c_{-i} represents the vector of functions $(\rho^c_{i1}, \dots, \rho^c_{i,i-1}, \rho^c_{i,i+1}, \dots, \rho^c_{in})$. The maximisation problem of player i becomes:

$$\max_{e_i} \left\{ V^i \left(\rho^c_{i1}(e_i; e^b), \dots, \rho^c_{i,i-1}(e_i; e^b), e_i, \rho^c_{i,i+1}(e_i; e^b), \dots, \rho^c_{in}(e_i; e^b) \right) \right\} .$$

Accordingly, the first order equation generalising (1.6) is:

$$V^i_i(e_i, e_{-i}) + \sum_{j \neq i} r_{ij}(e_i, e_{-i}) \, V^i_j(e_i, e_{-i}) = 0 . \tag{1.18}$$

A conjectural variations equilibrium is a strategy profile e^c which solves Equations (1.18) simultaneously for $i = 1, \dots, n$. The conjectural variation r_{ij} may depend on e_i, or both (e_i, e_{-i}).

[4] Here, the subscript "ij" stands for "player i's conjecture about player j".

Consistency in the case of many players

A conceptual difficulty arises when one considers consistency. The strongest notion of consistency requires that the conjectural best response of player i coincides with what other players have conjectured about her, that is, with one of their conjectured reaction functions. However, when n players are present, there are n best response functions χ_i and $n(n-1)$ conjectures r_{ij}, $i \neq j$. Therefore, if $n > 2$, an equilibrium is consistent only if all players have the same conjectures about player i. In formal terms: for each player i,

$$r_{ki}(e_k, e_{-k}) \;=\; r_{ji}(e_j, e_{-j})\;, \quad \forall j, k \neq i\;. \tag{1.19}$$

This is the approach followed explicitly in Olsder (1981). This assumption can be found also in Fershtman and Kamien (1985), for conjectures in differential games (see Section 3.2.2).

In the literature on static n-player conjectural variations, the problem is usually implicitly addressed by assuming a complete symmetry between agents. Perry (1982) for the oligopoly, or Cornes and Sandler (1984) and Sugden (1985) for public goods, consider a class of games where for each player, the contributions of all other players to her payoff are aggregated. It is as if each player plays against a unique (virtual) player representing the remaining players. The payoff functions under consideration are $V^i(e_i, y_i)$ where $y_i = \sum_{j\neq i} e_j$. The payoff of agent i thus depends on her strategies and the sum of the rival strategies. Each agent makes conjectures about how the other players, as a whole, react to her choices. It is assumed that the conjectural variations are all identical, *i.e.* $y_i = \rho_i^c(e_i)$, with $(\rho_i^c)' \equiv r$, $\forall i$.

In these papers, consistency is required only in the subspace of strategies where $e_i = e_j$ and punctually at the equilibrium. It is to be noted that this is not a natural extension of the definition of Bresnahan (1981) for two players, which naturally follows from Assumption (1.19). We shall nevertheless restrict our attention to this concept of consistency in the many-player illustrations of this monograph.

1.4 Examples and illustrations

This section develops the computation of conjectural equilibria in three classical examples: Cournot's and Bertrand's duopolies, and a model of voluntary contributions to a public good. We shall use these examples repeatedly as illustrations throughout the monograph. All these examples use the "standard" definition of conjectural variations equilibria as in Section 1.3.5, and the definition of consistency of Section 1.3.7. Furthermore, we only consider *constant* conjectural variations.

In Appendix A, we provide general results covering these cases. Our purpose for the time being is to illustrate step-by-step the different ideas.

1.4.1 *Cournot's duopoly*

Bresnahan (1981), Example 2 studies the case of a duopoly with perfectly substitutable goods, and with constant marginal costs. He shows that, under suitable assumptions on the inverse demand function, there is a constant CCVE with conjectural variation -1.

Consider the profit functions of firms in a Cournot duopoly with constant marginal cost c:

$$V^i(e_i, e_j) \;=\; p(e_i + e_j)\; e_i \;-\; c\, e_i\;.$$

Here, $p(\cdot)$ stands for the inverse demand function. It it assumed to be decreasing and concave, and admissible quantities are non-negative numbers such that $p(.)$ is positive.

Assume that the firms have identical linear conjectured reaction functions with slope r (that is, constant variational conjectures). Given a reference profile of quantities $e^b = (e^b_1, e^b_2)$, firm i assumes that

$$e_j \;=\; e^b_j \;+\; r\;(e_i - e^b_i)\;.$$

The optimisation problem of firm i (see Equation (1.3)) is:

$$\max_{e_i} \overline{V}^i(e_i)\;, \quad \text{with} \quad \overline{V}^i(e_i) = p(e_i + e^b_j + r(e_i - e^b_i))e_i - ce_i\;.$$

Since:

$$(\overline{V}^i)''(e_i) \;=\; (1+r)^2\; e_i\; p'' \;+\; 2\;(1+r)\; p'\;,$$

a sufficient condition for the existence and uniqueness of a solution is $1 + r > 0$.

In order to find the CVE corresponding to the conjecture r, one solves the system of Equations (1.10), that is:

$$(1+r)\; e_i\; p'(e_i + e_j) \;+\; p(e_i + e_j) \;-\; c \;=\; 0\;, \qquad i \neq j\;. \tag{1.20}$$

For instance, when the inverse demand function is linear with the form $p(E) = a - bE$, $a, b > 0$, the CVE is:

$$e_1^c = e_2^c = \frac{a-c}{b(3+r)} .$$

When $r = 0$, one recognises the Nash equilibrium of the game:

$$e_1^N = e_2^N = \frac{a-c}{3b} .$$

We now examine whether consistent conjectural equilibria exist for this game. According to Corollary 1.4, Equation (1.14), the constant r should be solution of the equation:

$$\begin{aligned} 0 &= (1+r^2)(e_i\, p'' + p') + r(e_i\, p'' + 2p') + r\, p'' \\ &= (1+r)^2\, (e_i\, p'' + p') . \end{aligned}$$

The only possible value for r is -1. In order to see that $r = -1$ and some strategy profile form indeed a CCVE, it is necessary to check that the corresponding equilibrium value maximises the perceived profit of firms. Any equilibrium (e_1^{cc}, e_2^{cc}) is solution of (1.20), that is, for $r = -1$,

$$p(e_1^{cc} + e_2^{cc}) = c .$$

In other words, the price at the equilibrium is equal to the marginal cost. In that case, the perceived profit $\overline{V}^i(e_i)$ becomes:

$$\begin{aligned} \overline{V}^i(e_i) &= e_i\, p(e_i + e_j^{cc} - (e_i - e_i^{cc}))e_i - c\, e_i \\ &= e_i\, \left(p(e_i^{cc} + e_j^{cc}) - c\right) \\ &= 0 . \end{aligned}$$

Therefore, *any* couple of strategies (e_1^{cc}, e_2^{cc}) such that $p(e_i^{cc} + e_j^{cc}) = c$, and the conjectural variations $r_1 = r_2 = -1$ form indeed a CCVE. Among them, the symmetric profile corresponds to Bertrand's equilibrium. As noted by Bresnahan (1981), the fact that Bertrand's equilibrium be a *consistent* conjectural equilibrium depends critically on the fact that marginal costs of this duopoly be constant. In Chapter 2, we come back to this model in a dynamic setting. In Section C.1 of the Appendix, we compare profits obtained with the Cournot-Nash equilibrium, conjectural equilibria, and the cartel solution.

1.4.2 *Bertrand's duopoly*

Sterdyniak and Villa (1993) study the conjectural equilibria with a constant conjecture for a duopoly model with the price as the strategic variable: Bertrand's duopoly. The profit functions are of the from:

$$V^i(e_i, e_j) = (e_i - A)(y_0 - \alpha e_i + \beta e_j) - B \ , \tag{1.21}$$

e_i being the price set by firm i, A the marginal cost, and B the fixed cost. According to Sterdyniak and Villa (1993), the cases of economic relevance are: $0 < \beta < \alpha$ (substitutable goods) and $\beta < 0 < \alpha$ (complementary goods).

For a conjectural variations equilibrium with a constant variational conjecture equal to r (which is supposed to be the same for both firms, given the symmetry of the profit function), one obtains from Equation (1.10) the following first order conditions:

$$V_i^i(e_i, e_j) + r\ V_j^i(e_i, e_j) = y_0 + \beta e_j + (\alpha - r\beta)A + (-2\alpha + r\beta)e_i = 0 \ ,$$

$i, j = 1, 2$. The solution is:

$$e_1^c = e_2^c = \frac{y_0 + (\alpha - r\beta)A}{2\alpha - (1+r)\beta} \ . \tag{1.22}$$

In order to check the maximality condition, consider:

$$\begin{aligned} \overline{V}^i(e_i) &= V^i(e_i, e_j^b + r(e_i - e_i^b)) \\ &= (e_i - A)(y_0 - \alpha e_i + \beta(e_j^b + r(e_i - e_i^b))) - B \ . \end{aligned}$$

This gives:

$$(\overline{V}^i)''(e_i) = -2(\alpha - \beta r) \ .$$

In order to obtain a maximum, it is necessary to consider $r < \alpha/\beta$ if $\beta > 0$ and $r > \alpha/\beta$ if $\beta < 0$. Note that this condition eliminates the possibility that the denominator of (1.22) is equal to zero.

Finally, in the particular case $r = 0$ one obtains the Nash equilibrium which, in this case, is the Bertrand equilibrium:

$$e_1^N = e_2^N = \frac{y_0 + \alpha A}{2\alpha - \beta} \ .$$

It is easy to check that this static Nash equilibrium is not consistent in the sense of Bresnahan. For instance, the necessary condition of Theorem A.7, page 132, fails to be satisfied, since:

$$V_1^1(e_1^N, e_2) \;=\; \beta \neq 0 \qquad \forall e_2 \in E_2\;.$$

For constant consistent conjectures, Corollary 1.4 implies that r must satisfy the equation:

$$\beta r^2 - 2\alpha r + \beta = 0\;, \tag{1.23}$$

which gives, provided that $|\alpha| \geq |\beta|$, a single value which satisfies the restriction $\alpha - \beta r > 0$ required above:

$$r \;=\; \frac{1}{\beta}\left(\alpha - \sqrt{\alpha^2 - \beta^2}\right)\;.$$

Notice that β and r have the same sign. Denote by e_i^{cc} the consistent conjectural equilibrium obtained by replacing r by this value in (1.22). Since the profit function $\overline{V}^i(e_i)$ is increasing with respect to βr, we obtain:

$$V^i(e_1^N, e_2^N) < V^i(e_1^{cc}, e_2^{cc}) \text{ if } \quad \beta > 0$$

$$V^i(e_1^{cc}, e_2^{cc}) < V^i(e_1^N, e_2^N) \text{ if } \quad \beta < 0\;.$$

A general framework for the comparison of payoffs obtained with the different solution concepts is described in Appendix B.

1.4.3 *Voluntary contributions to a public good*

In this section, we discuss an example of conjectural variations equilibria in models of voluntary contributions to a public good, as for instance in Itaya and Dasgupta (1995), but in a symmetric context.

Consider two agents, who may choose to contribute to a public good G or spend in a private good. Given a total budget I, let e_i be the contribution to the public good, and $I - e_i$ the amount devoted to the private good consumption. The public good is produced using a linear technology, so that $G = e_1 + e_2$. Assume further that individual's preferences are represented by Cobb-Douglas utility functions:

$$V^i(e_i, e_j) = (I - e_i)^\alpha (e_i + e_j)^{1-\alpha}$$

for $i = 1, 2$, with the condition $\alpha < 1/2$.

Assume finally that agents form constant variational conjectures r.

Given a benchmark strategy profile e^b, the optimisation problem of agent i (see Equation (1.3)) is:

$$\max_{e_i} \overline{V}^i(e_i) \ , \quad \text{with} \quad \overline{V}^i \ = \ (I - e_i)^\alpha (e_i + e_j^b + r(e_i - e_i^b))^{1-\alpha} \ .$$

Since, calling $A_i = e_i + e_j^b + r(e_i - e_i^b)$:

$$\begin{aligned}(\overline{V}^i)''(e_i) = \alpha(\alpha-1)(I-e_i)^{\alpha-2}A_i^{1-\alpha} \\ -2\alpha(\alpha-1)(I-e_i)^{\alpha-1}(1-\alpha)(1+r)A_i^{-\alpha} \\ -\alpha(\alpha-1)(I-e_i)^{\alpha}(1-\alpha)(1+r)^2A_i^{-\alpha-1} \ ,\end{aligned}$$

a sufficient condition for the existence and uniqueness of a solution is

$$1 + r \ > \ 0 \ .$$

To obtain the CVE, we write down Equation (1.10), which in this case gives, after simplifications:

$$-\alpha \ (e_1 + e_2) \ + \ (1-\alpha) \ (1+r) \ (I - e_i) \ = \ 0 \ , \qquad (1.24)$$

for $i \neq j$. Solving these equations gives the unique CVE:

$$e_1^c = e_2^c \ = \ I \ \frac{(1-\alpha)(1+r)}{(1-\alpha)(1+r)+2\alpha} \ .$$

In order to find consistent CVE, one may compute the conjectural best response function of agent i, solving Equation (1.24) for e_i as a function of e_j:

$$\chi_i(e_j) \ = \ \frac{I(1-\alpha)(1+r) - \alpha e_j}{(1-\alpha)(1+r) - \alpha} \ .$$

Then, the CVE is consistent if and only if $\mathrm{d}\chi_i(e_j)/\mathrm{d}e_j = r$, which amounts to requiring that:

$$r \ = \ - \ \frac{\alpha}{(1-\alpha)(1+r)-\alpha} \ .$$

There are two solutions of this equation:

$$r \ = \ -1 \ , \qquad r \ = \ - \ \frac{\alpha}{1-\alpha} \ .$$

It can be checked that the solution corresponding to $r = -1$ is Pareto-dominated by the other. The reasonable CCVE is therefore

$$e_1^{cc} = e_2^{cc} \ = \ I \ (1 - 2\alpha) \ .$$

We come back to this model in Section C.2 of Appendix C, where we discuss the comparison of different conjectural equilibria in terms of efficiency.

1.5 An attempt at generalisation

So far the introduction of the conjectures was made through differential equations; no more, no less was needed to investigate how conjectures play a role *via* the way they distort optimal decisions, at the level of first order conditions. We try in this section to shift the emphasis from first order conditions to the conjectures themselves. This will produce a more general presentation of conjectural equilibria (inspired by Laitner (1980), Başar and Olsder (1999), §4.5 and Friedman (2003)) for, at least in principle, it becomes possible to consider the role of conjectures on behaviours even if the decisions are not *continuous* variables.

Consider the point of view of player i. Assume that to each benchmark strategy profile $e^b \in \mathbb{E}$, player i associates a subset $\mathcal{C}^i_{e^b}$ of $\mathbb{E}$ such that $e^b \in \mathcal{C}^i_{e^b}$.

The interpretation of $\mathcal{C}^i_{e^b}$ is the following. Taking the particular strategy profile e^b as a reference, player i assumes that if she modifies her own strategy (not necessarily in an infinitesimal way) into e^m_i, then player j's strategy will be to play some e^m_j such that the profile (e^m_1, e^m_2) will stay in the same set $\mathcal{C}^i_{e^b}$. The set $\mathcal{C}^i_{e^b}$ is the *conjectured reaction set* for player i and for the reference strategy e^b. As we shall see below, the family of sets $\{\mathcal{C}^i_{e^b}\}$ should have the *equivalence property* that if $f \in \mathcal{C}^i_e$, then $\mathcal{C}^i_e = \mathcal{C}^i_f$. In other words, if f is expected to be a reasonable deviation from e, then all reasonable deviations from f are also reasonable deviation from e.

In the situations we have encountered in this chapter, $\mathcal{C}^i_{e^b}$ is normally a simple curve in the plane, or the graph of a function. Since in addition those functions are solution to ordinary differential equations, they enjoy the equivalence property. However, in general it is not necessary to assume that the expected change of the opponent associated with some deviation should be unique.

Given this belief about her opponent, player i maximises her payoff by choosing the (optimal) reference strategy *in the set* $\mathcal{C}^i_{e^b}$. This defines, for each value of e^b, a set of optimal strategy profiles:

$$\phi_i(e^b) = \arg \max_{(e_1,e_2)\in\mathcal{C}^i_{e^b}} V^i(e_1,e_2) . \tag{1.25}$$

It is worth stressing that $\phi_i(e^b)$ is a *set* of two-dimensional vectors. Naturally, in regular cases, the optimisation problem (1.25) is expected to have an unique solution, and the set $\phi_i(e^b)$ will be reduced to a single strategy profile. Also, this set may be empty if the maximisation problem does not have a solution in $\mathcal{C}^i_{e^b}$.

Finally, considering all the admissible benchmark profiles, the union $\mathcal{R}^i = \cup_{e^b} \phi_i(e^b)$ constitutes the *conjectural best response set* of player i associated with her family of conjectures $\{\mathcal{C}^i_{e^b}\}$. This terminology comes from the observation that every strategy profile in the set $\mathcal{R}^i$ is indeed the best player i can do "in response" to some exogenously given benchmark strategy. The name is further justified by the fact that this set coincides with the familiar Nash best response function when conjectures are constant. See the discussion in Section 1.3.4.

A strategy profile is then an equilibrium if no player thinks it is in her interest to deviate from it, given her conjectures. If any player adopts this profile as benchmark, she will find no interest to play another strategy. In other words, any equilibrium e^c should be such that:

$$e^c \in \phi_1(e^c) \cap \phi_2(e^c) . \tag{1.26}$$

Such a strategy must also be in the intersection of both players' conjectural best response sets, just as in the usual case of Nash best responses.

When the set $\mathcal{R}^i$ contains one single curve, the set of conjectural equilibria is then $\mathcal{R} := \mathcal{R}^1 \cap \mathcal{R}^2$. When the sets $\mathcal{R}^i$ are made of multiple curves, the set of conjectural equilibria is obtained from the intersection(s) of the curves that belong to the sets $\mathcal{R}^1$ and $\mathcal{R}^2$.

Consider now the notion of *consistency* of some conjectural equilibrium (e^c_1, e^c_2). From the point of view of player i, the reasoning is the following. If she chooses to deviate from the equilibrium value $e^c = (e^c_i, e^c_j)$ to some e^m_i, she believes that, according to her set of conjectures, player j should react and play some value e^m_j such that $(e^m_i, e^m_j) \in \mathcal{C}^i_{e^c}$. But in fact, player j would react, according to her own computations, by playing a conjectural best response strategy e^{br}_j corresponding to some strategy profile in the set $\phi_j(e^m)$. Therefore, there will be a mismatch between player i's predictions and the observed behaviour unless $\phi_j(e^m) \subset \mathcal{C}^i_{e^c}$. Since this coincidence must hold for any deviations e^m_i, it is necessary that all the best response set $\mathcal{R}^j$ of player j be included in the conjecture $\mathcal{C}^i_{e^c}$ of player i.

We summarise the concepts in the following definitions.

Definition 1.8 (CONJECTURE AND CONJECTURED REACTION SET) A *conjecture* formed by player i is a family of *conjectured reaction sets* $\{\mathcal{C}^i_{e^b}; e^b \in \mathbb{E}\}$. This family is such that $e^b \in \mathcal{C}^i_{e^b}$, for all e^b, and the *equivalence property* holds.

Definition 1.9 (CONJECTURAL BEST RESPONSE) The *conjectural best response* of player i, given this conjecture, is the set of best responses $\mathcal{R}^i$.

Definition 1.10 (CONJECTURAL EQUILIBRIUM) A *conjectural equilibrium* corresponding to this conjecture is any strategy in the intersection of both players' conjectural best response sets: $\mathcal{R} = \mathcal{R}^1 \cap \mathcal{R}^2$.

Definition 1.11 (CONSISTENT CONJECTURAL EQUILIBRIUM) A conjectural equilibrium $e^c = (e^c_1, e^c_2)$ corresponding to a pair of conjectures $(\mathcal{C}^1_{e^c}, \mathcal{C}^2_{e^c})$ is *consistent* if the corresponding reaction sets $\mathcal{R}^1, \mathcal{R}^2$ satisfy:

$$\mathcal{R}^i \subset \mathcal{C}^j_{e^c}\ , \qquad i,j = 1,2\ , \qquad i \neq j\ .$$

An important point to check is that equilibria, as defined by (1.26) are indeed the same as the intersection of best response sets. We have the:

Theorem 1.6 *Let*

$$\mathcal{E} = \{e^c \in \mathbb{E} \mid e^c \in \phi_1(e^c) \cap \phi_2(e^c)\}\ .$$

The sets $\mathcal{E}$ and $\mathcal{R}$ coincide.

Proof. Take $e \in \mathcal{E}$. Since $e \in \phi_i(e) \subset \mathcal{R}^i$, for all i, then $e \in \cap_i \mathcal{R}^i = \mathcal{R}$. Therefore, $\mathcal{E} \subset \mathcal{R}$.

Now, let $e \in \mathcal{R}$. For each i, there exists a strategy profile f^i such that $e \in \phi_i(f^i)$. We shall prove that actually, $e \in \phi_i(e)$. By definition, $\phi_i(f^i) \subset \mathcal{C}^i_{f^i}$. Therefore, $e \in \mathcal{C}^i_{f^i}$, which implies, by the equivalence property of conjectured reaction sets, that $\mathcal{C}^i_e = \mathcal{C}^i_{f^i}$. This, in turn, implies that the conjectural optimisation problems in (1.25), are the same with e or f^i as benchmark. We conclude that: $\phi_i(e) = \phi_i(f^i)$. In summary, we have obtained that $e \in \phi_i(f^i) = \phi_i(e)$ for all i, hence $e \in \cap_i \phi_i(e)$, and $e \in \mathcal{E}$. □

We conclude this section by showing that the concepts of conjectural *variations* discussed in this chapter are indeed particular cases of this formalism.

Going back to Paragraph 1.3.4.1, we see that for any benchmark strategy e^b, we have defined, for player i, a conjectured reaction *function*

$$e_j = \rho^c_j(e_i; e^b)$$

by solving the ordinary differential equation (1.2) with the initial condition $e_j^b = \rho_j^c(e_i^b; e^b)$. In the terminology of this section, the conjectured reaction set $\mathcal{C}^i_{e^b}$ is a simple curve: the graph of this function. In other words, this is the set of strategy profiles of the form

$$\mathcal{C}^i_{e^b} = \{(e_i, e_j) \in \mathbb{E}, \quad e_j = \rho_j^c(e_i; e^b)\} .$$

We have assumed that the differential equation (1.2) is regular enough to have a unique solution. Under this condition, any point on the curve can be taken as the initial condition of the same differential equation, and lead to the same solution. This means that such conjectured reaction sets automatically enjoy the *equivalence property.*

As another illustration of the concepts, consider again the "Nash play". In this situation, players conjecture that their opponent will play a fixed strategy independent of their own. Still in the terminology above, this corresponds to a conjecture formed of the conjectured reaction sets

$$\mathcal{C}^1_{(e_1, e_2)} = \{(e, e_2), e \in E_1\}$$

(and symmetrically for player 2), for every reference strategy $(e_1, e_2) \in \mathbb{E}$. The reaction set $\mathcal{R}^i$ is then the graph of the familiar Nash best response function $\chi_i^N(e_j) = \arg\max_{e_i} V^i(e_i, e_j)$.

1.6 Conclusion

This first chapter has revisited the quite old concept of CVE in the static case, given that this concept is mentioned in recent texts as a possible complement or extension of the original Nash's equilibrium (see *e.g.* Başar and Olsder (1999)). We have made an attempt at clarifying the notion of conjectural variations equilibrium, by revisiting formally several works on the topic, classifying them according to the type of conjectured reaction and strength of consistency, and discussing their relationship. We have adopted a quite strict formalism. Although this approach obviously leads to cumbersome formulas, we believe it proves fruitful, in helping avoid confusions and errors.

Our next step is to develop, in the following chapters, the linkage between conjectural variations equilibria and dynamic games. Several authors of papers on CVE have commented on the necessity of expressing conjectures in a dynamic context. For instance, the conclusion of Fershtman and Kamien (1985) is: *It is clear to us that the use of conjectural variations is essential to the understanding of dynamic interaction among players and we are convinced that the controversy of it can be resolved within the framework of dynamic games [...].*

Also, the survey of CVE presented in this chapter is by no means complete, and we come back to more technical developments in the appendix. We shall successively: identify geometric relationships between consistent equilibria and iso-payoff curves, and examine cases where CCVE exist (Appendix A), then expose results on the comparison of conjectural variations equilibria, Nash equilibria and Pareto outcomes in terms of efficiency (Appendix B). The classical examples of the literature are revisited in Appendix C.

Chapter 2

Conjectures as Reduced Forms for Dynamic Interactions

2.1 Introduction

This chapter explains how it is possible to interpret the conjectural variations equilibrium (CVE), for some static games, as a reduced form for dynamic competition. This kind of reinterpretation has been proposed by several authors in two archetypical economic situations, in the context of the private provision of a public good (Itaya and Shimomura (2001); Itaya and Okamura (2003)) and in the context of the oligopoly (Dockner (1992); Driskill and McCafferty (1989); Cabral (1995)). As this chapter shows, the same kind of reinterpretation can be followed in the context of public infrastructures competition (Wildasin (1991); Figuières (2002)). A first interest is to provide an acceptable theoretical justification for the controversial static CVE model; but the other way around the CVE might provide a useful shortcut for more sophisticated dynamic interactions.

When interactions repeat over time, strategies become very much more complex to analyse. Two kinds of strategy spaces are widely used in Dynamic Games, the open-loop strategy space and the feedback strategy space. In the first case, each player takes the rival strategies as simple functions of time when she determines the optimal trade-off between the current and the future effects of her actions. In the second case those intertemporal trade-offs are modified because each player considers rival actions at any period to be functions of the state of the game at the same period; as a consequence every player takes into account the modifications of rival actions after the change she induces to the state variable. We shall also consider a third possible mood of play, involving trigger strategies: players stick to a given profile of actions unless a deviation by any player is observed, which then triggers punishment actions by the opponents. Such strategies formally capture the ideas of retaliations and threats.

Note that these three dynamic behavioural characteristics have no natural place in static games. When applied to a given context, they generally give rise to outcomes that differ from the static Nash equilibrium. However, as this chapter explains, there exists some relationship between the dynamic equilibrium concepts and the static conjectural variations equilibrium. We present slightly simplified versions of the seminal references, which are sufficient to figure out the arguments; readers are invited of course to go through the original papers.

The chapter is organised as follows. Section 2.2 studies the classical problem of private provisions to a public good. It successively presents the static, repeated and dynamic versions of the voluntary contributions model. Section 2.3 deals with the static, repeated and dynamic versions of the oligopoly. Then Section 2.4 considers a problem of public infrastructure competition, both from static and dynamic points of view. In all these economic situations it is shown how the static CVE, to some extent, captures more complex dynamic phenomena. Section 2.5 is devoted to a more general class of capital accumulation games, which encompasses all the other dynamic games introduced in the chapter, and for which we analyse the link between the payoff structure and the nature of the conjecture of the associated static game. Section 2.6 summarises the chapter. Technical proofs and computations are provided in Section 2.7.

2.2 Private provision of a public good

2.2.1 *One-shot simultaneous contributions*

Recall the public good framework outlined in Chapter 1. In a simple version there are n identical agents who are all endowed with the same exogenous revenue I. Agent i uses this revenue to consume a single private good c_i, which price has been normalised to unity, and she contributes e_i to a public good G. This public good is the sum of all the contributions, *i.e.* $G = \sum_{i=1}^{n} e_i$. Choosing units such that the relative price between the two goods is equal to one, her budget constraint is simply:

$$I = c_i + e_i \ .$$

The utility function of each agent depends on two arguments: the consumptions of the private good and of the public good. It is denoted $U^i(c_i, G)$ and it is assumed concave with positive partial derivatives $U^i_c \geq 0, U^i_G \geq 0$.

Any agent i is confronted with a trade-off between the consumption of the private good and the contribution to the public good. How much will she decide to contribute? A prediction can be given when one sees the individual decisions as the outcome of a noncooperative game. If $G_{-i} = \sum_{j \neq i} e_j$ stands for the others' contributions, substituting the budget constraint into the utility function, and assuming that agents conjecture the existence of a relationship $G^c_{-i}(e_i)$ between their decisions and the sum of the others' decisions, the decision problem of each agent is:

$$\max_{e_i \geq 0} U^i \left(I - e_i, e_i + G^c_{-i}(e_i)\right) .$$

The first order conditions for an interior conjectural variations equilibrium are:

$$U^i_c(I - e_i, G) \;=\; (1 + r)\, U^i_G(I - e_i, G) \qquad \forall i \, , \tag{2.1}$$

where

$$r \;=\; \frac{\mathrm{d}G^c_{-i}}{\mathrm{d}e_i} \, .$$

The first order conditions boil down to the first order conditions for an interior Nash equilibrium only when $r = 0$.

The aggregate Nash supply of the public good generally falls short of the Pareto optimal level as formally defined by the Samuelson rule (see *e.g.* Laffont (1988); Cornes and Sandler (1996)). Restricting attention to agents with identical utility functions for simplicity, the Samuelson rule can be written:

$$U^i_c(I - e_i, G) \;=\; nU^i_G(I - e_i, G) \qquad \forall i \, . \tag{2.2}$$

The intuition is straightforward. Comparing (2.2) and (2.1) when $r = 0$, it follows that the marginal individual benefits of the public good are n times less important than the social marginal benefit. As a result, noncooperative incentives to contribute are too weak. On one hand, each individual benefits freely from others' contributions, and on the other hand she does not take into account the benefit her contribution brings to the others' utilities; this is a free rider problem.

With positive conjectures ($r > 0$) individual incentives to invest increase, and so does the supply of the public good. Clearly, efficiency could be implemented on a noncooperative basis when the conjecture takes the particular value $r = n - 1$. A positive but less than $n - 1$ conjecture will increase the supply of the public good, which is a Pareto improvement with respect to the Nash equilibrium in this context of positive externalities (see also Section B.2, page 148). On the contrary a negative conjecture will cut the marginal benefits of the public good, reduce the supply and will bring about a Pareto deterioration.

If static conjectural games are a reduced form for intertemporal games it is then interesting to study whether dynamic interactions soften or increase the free rider problem (r positive or negative).

2.2.2 *Repeated contributions*

What if the above setting is repeated *ad infinitum*? To study this question Itaya and Okamura (2003) focus on two specific families of examples, with quadratic and Cobb-Douglas utility functions. We shall report only the first case here where preferences are prescribed as:

$$U^i(c_i, G) = c_i + G(A - G) \ , \quad G \leq A/2 \ .$$

A is a preference parameter; the higher this parameter the higher the marginal utility of the public good for a given level G.

In this quadratic case, the first order conditions for the static CVE (2.1) can be rewritten as:

$$e_i^c = \frac{1}{2}\left(A - \frac{1}{1+r}\right) - \sum_{j \neq i} e_j^c \ , \quad i = 1, \dots, n \ . \tag{2.3}$$

Now consider that the economic situation at hand is seen as a repeated game of complete information where player i's payoff is the discounted stream of the above per-period utilities, with discount factor $\theta \in (0, 1)$.

Strategies are then typically more complex as players can condition their current decision upon the whole history of past moves. From the Folk Theorem it is known that many outcomes can be supported as Nash equilibria, including full or partial cooperation depending on how much players discount the future. From the point of view of an outside observer who would consider only the situation at period t, the outcome differs from the static Nash equilibrium. Rather, as shown below, it is 'observationally' equivalent to a static conjectural variations equilibrium with positive conjectures. Interestingly the conjectural variations that appear here are endogenously determined from inter-temporal optimising behaviours; it is possible to relate these conjectures to the structural parameters of the model and to see, for instance, if the conjectural variation increases with the discount factor θ or the marginal utility for the public good (when A increases).

A given prescribed profile of contributions $(e_1^*, \ldots, e_n^*)$, possibly including fully efficient contributions, can be supported at each period as a Nash equilibrium when players pick trigger strategies. Such strategies work as follows. In any period each player sticks to e_i^* provided that all players actually chose their prescribed level in the previous period. Otherwise if player j alone deviated she is punished in the current period: she receives the harshest possible punishment, for any other player will contribute $e_i = 0$. After one period of punishment the prescribed levels are played again.

Let us check that such a profile of strategies can indeed be a Nash equilibrium under appropriate conditions.[1] Clearly only the payoffs of the deviating period and the consecutive period of punishment need to be investigated and compared with the payoff of the same periods when players choose the prescribed levels.

The highest current level of utility a player can enjoy from deviation while the other players stick to their prescribed levels is:

$$\widehat{U}^i = U^i\left(I - \widehat{e}_i, \widehat{e}_i + E_{-i}^*\right) ,$$

where $\widehat{e}_i$ is a best response to $E_{-i}^* = \sum_{j \neq i} e_j^*$. Computing $\widehat{e}_i$ in our quadratic framework and plugging the result back into the utility function, one has:

$$\widehat{U}^i = I + \sum_{j \neq i} e_j^* + \frac{1}{4}(A-1)^2 .$$

Similarly the highest per-period level of utility a punished player can enjoy is:

$$\overline{U}^i = U^i\left(I - \overline{e}_i, \overline{e}_i\right) ,$$

where $\overline{e}_i$ is a best response to zero contributions by the others. This best response can easily be calculated and substituted into the utility function to give:

$$\overline{U}^i = I + \frac{1}{4}(A-1)^2 .$$

[1] It is worth noting that this equilibrium is not subgame perfect. Indeed if a player does deviate, it is not in the best interest of the others to actually implement their threats. The trigger strategies that will be used later in this chapter for the repeated oligopoly share the same limitation. Restricting attention to subgame perfect trigger strategies would be interesting extensions of Itaya and Okamura (2003) and Cabral (1995).

Finally let us denote

$$U^{i*} = U^i\left(I - e_i^*, e_i^* + E_{-i}^*\right) ,$$

the level of per-period utility resulting from the prescribed contributions.

For the prescribed levels to form a Nash equilibrium it is necessary that, for every player, the resulting payoff over the current and the next period is at least equal to the sum of the payoff associated with the deviation and the discounted payoff associated with the subsequent punishment. Formally:

$$(1+\theta)\, U^{i*} \geq \widehat{U}^i + \theta \overline{U}^i ,$$

or more precisely in the quadratic payoff case:

$$(1+\theta)\left[I - e_i^* + G^*\left(A - G^*\right)\right] \geq I + E_{-i}^* + \frac{1}{4}(A-1)^2 + \theta\left[I + \frac{1}{4}\left(A-1\right)^2\right]$$

Using the fact that $G^* = e_i^* + E_{-i}^*$ and rearranging this last inequality one has:

$$\left(e_i^*\right)^2 - \left(A - 1 - 2E_{-i}^*\right) e_i^* + \left[E_{-i}^* - \frac{1}{2}\left(A-1\right)\right]^2 - \frac{\theta}{1+\theta} E_{-i}^* \leq 0 .$$

Solving this inequality for e_i^* :

$$\frac{1}{2}\left(A-1\right) - E_{-i}^* - \sqrt{\frac{\theta}{1+\theta} E_{-i}^*} \leq e_i^* \leq \frac{1}{2}\left(A-1\right) - E_{-i}^* + \sqrt{\frac{\theta}{1+\theta} E_{-i}^*} .$$

It follows that the maximum attainable level of prescribed contribution is:

$$e_i^* = \frac{1}{2}\left(A-1\right) - E_{-i}^* + \sqrt{\frac{\theta}{1+\theta} E_{-i}^*}. \tag{2.4}$$

At a symmetric equilibrium where $e_i^* = e_j^* = e^*$, this equation reads as:

$$ne^* = \frac{1}{2}\left(A-1\right) + \sqrt{\frac{\theta}{1+\theta}\left(n-1\right) e^*} .$$

This expression is a quadratic equation in $\sqrt{e^*}$. Solving this equation one finds

$$e^* = \frac{1}{4n^2}\left[\varphi + \sqrt{\varphi^2 + 2n\left(A-1\right)}\right]^2 ,$$

where $\varphi = \sqrt{\frac{\theta(n-1)}{1+\theta}}$. Notice that the solution e^* is an increasing function of the marginal payoff for the public good (as captured by the parameter

A) and of the discount factor θ. At one extreme when $\theta \to 0$, then

$$\lim_{\theta \to 0} e^* = \frac{1}{2n} (A - 1) ,$$

which is the static Nash equilibrium. Therefore as θ increases the maximum sustainable contribution (without the risk of a deviation from a rival) leaves this benchmark Nash level and goes up towards the Pareto level.

What is the relationship between e^* and the corresponding symmetric static CVE? If $e^c = e^*$, then from (2.3) and (2.4) it follows:

$$\frac{r}{1+r} = 2 \sqrt{\frac{\theta}{1+\theta}} (n-1) e^* ; \tag{2.5}$$

this last expression indicates the implicit relationship between the static constant conjecture on one hand and a given contribution level e^* in the repeated game and the discount factor on the other hand. It is readily checked that the higher the marginal payoff for the public good (*i.e.* the higher A) the higher e^* and therefore the higher the conjecture r. Also the higher the discount factor θ and the higher the conjecture.

What is the specific value of the discount factor for which the maximum level is compatible with the Pareto-optimal level? Using the Samuelson condition the efficient level of public good and individual contributions can be found to be:

$$G^P = \frac{1}{2} \left(A - \frac{1}{n} \right) , \qquad e^P = \frac{G^P}{n} .$$

Compatibility of e^* with the Pareto-efficient level then requires that:

$$e^P = \frac{1}{2} (A - 1) - (n-1) e^P + \sqrt{\frac{\theta}{1+\theta}} (n-1) e^P .$$

Solving this equation for θ:

$$\theta^* = \frac{n-1}{(2A-1) n - 1} ,$$

which is the lowest value for the discount factor that allows the Pareto-optimal level of public good to be sustained as a Nash equilibrium in trigger strategies of the repeated game.

What conclusion, if any, can be drawn as for the conjectural parameter? First, there exists a value r such that the CVE with constant variations r is the same as the optimal strategy profile of the repeated game. Next, from the implicit relationship (2.5), if this parameter is meant to capture the above underlying inter-temporal threats, then: *i*) it has to be non negative, *ii*) it must be increasing with the marginal utility of the public good, for e^* increases with A, *iii*) and increasing with the discount factor.

2.2.3 *Private investment in a stock of public good*

Fershtman and Nitzan (1991) and Itaya and Shimomura (2001) have revisited this problem within a dynamic state-space framework. The n contributors are endowed with an exogenous revenue I at each point of time $t \in [0, +\infty)$. They use this revenue to consume a single private good c_i and to invest e_i to form a public good stock that accumulates according to the law of motion:

$$\dot{G} = \sum_{j=1}^{n} e_j - \delta G \,, \quad G(0) = G_0 \text{ exogenously given,} \tag{2.6}$$

where $\delta \geq 0$ is the physical rate of depreciation.

As before the current-period utility function of each agent depends on c_i and G. Let us assume more precisely that the utility function takes the quadratic form:

$$U^i(c_i, G) = \alpha + \beta_1 c_i - \frac{\gamma_1}{2}(c_i)^2 + \beta_2 G - \frac{\gamma_2}{2}(G)^2 \;,\; \alpha, \beta_1, \gamma_1, \beta_2, \gamma_2 > 0.$$

Finally, agent i's intertemporal discounted utility is:

$$\int_0^{\infty} e^{-\theta t} U^i(c_i(t), G(t))\ dt \,, \tag{2.7}$$

where $\theta > 0$ denotes the common discount rate.

Itaya and Shimomura (2001) propose to investigate the relationship between the conjectural equilibria of the static game and the steady state equilibria of the dynamic contribution game when the rate of time preference and of depreciation are sufficiently close to zero and one respectively.

Let us characterise the possible equilibrium behaviours in the dynamic game, starting first with the case where contributors commit to path strategies, leading to the so-called open-loop Nash equilibrium (OLNE). Each agent i maximises her intertemporal payoff (2.7) subject to (2.6) and given $e_j(t)$, $j \neq i$. At an OLNE each strategy must satisfy Pontryagin's Maximum Principle. If μ_i stands for agent i's costate variable, the current-value

Hamiltonian for agent i is

$$H^i = U^i(I - e_i, G) + \mu_i \left[\sum_{j=1}^{n} e_j - \delta G \right] .$$

The first order conditions for an interior solution are:

$$\begin{aligned} H_c^i &= 0 \Longrightarrow U_c^i = \mu_i , \quad \forall i , \\ \dot{\mu}_i &= \theta \mu_i - H_G^i \Longrightarrow \dot{\mu}_i = (\theta + \delta)\mu_i - U_G^i , \quad \forall i . \end{aligned} \tag{2.8}$$

At a steady state these conditions simplify to:

$$\frac{U_G^i}{U_c^i} = \theta + \delta , \quad \forall i.$$

When θ and δ are sufficiently close to zero and one respectively, these first order conditions almost coincide with those of the Nash equilibrium in the static game.

With the second equilibrium concept, the feedback Nash equilibrium (FBNE), agents select their optimal contribution decisions in every possible subgames. Let $W(G)$ denote the individual value function of this symmetric game: (2.7) started with stock G and evaluated at the equilibrium. The symmetric feedback Nash strategies must satisfy the single Hamilton-Jacobi-Bellman (HJB) equation:

$$\theta W(G) = \max_{e_i} \left\{ U^i(I - e_i, G) + W'(G) \left[\sum_{j \neq i} e_j(G) + e_i - \delta G \right] \right\} . \tag{2.9}$$

Any interior solution to the maximisation of the right-hand side of (2.9) satisfies:

$$U_c^i = W' . \tag{2.10}$$

For linear-quadratic games it is well know that a quadratic value function solves the Hamilton-Jacobi-Bellman equation (see for instance Başar and Olsder (1999)). Then it follows from the first order condition (2.10) that the FBNE strategies are linear functions of the stock of public good reached at each point in time, *i.e.*

$$e_i(G) = \phi_1 + \phi_2 G . \tag{2.11}$$

For a derivation of the convergent[2] FBNE see Section 2.7.1, where it is also shown that:

$$\phi_2 = \frac{1}{2n-1}\left(\frac{\theta}{2}+\delta-\sqrt{\left(\frac{\theta}{2}+\delta\right)^2+\frac{\gamma_2}{\gamma_1}(2n-1)}\right) < 0 .$$

Let us now interpret the steady state of the dynamic game as the outcome of a reduced-form static conjectural game. At the steady state $\overline{G}$ of the public good, the individual contributions are $\overline{e}_i$, with $\overline{G}=\sum_i \overline{e}_i$. The following relation holds:

$$\delta\overline{G}_{-i} = \sum_{j\neq i} e_j\left(\overline{e}_i+\overline{G}_{-i}\right)+(1-\delta)\,\overline{e}_i .$$

Total differentiation with respect to $\overline{G}_{-i}$ and $\overline{e}_i$ yields at a symmetric outcome:

$$\delta\mathrm{d}\overline{G}_{-i} = (n-1)\,e'\left(\overline{G}\right)\mathrm{d}\overline{e}_i+(n-1)\,e'\left(\overline{G}\right)\mathrm{d}\overline{G}_{-i}+(1-\delta)\,\mathrm{d}\overline{e}_i ,$$

therefore

$$\frac{\mathrm{d}\overline{G}_{-i}}{\mathrm{d}\overline{e}_i} = \frac{1-\delta+(n-1)\,e'\left(\overline{G}\right)}{\delta-(n-1)\,e'\left(\overline{G}\right)} , \qquad i=1,\ldots,n. \tag{2.12}$$

Substituting the linear strategies (2.11) into (2.12) gives the steady state conjectural variation:

$$\frac{\mathrm{d}\overline{G}_{-i}}{\mathrm{d}\overline{e}_i} = \frac{1-\delta+(n-1)\phi_2}{\delta-(n-1)\phi_2} = r.$$

The sign of this conjectural variation depends on the sign of the numerator, and it is indeterminate. Clearly it will be negative if δ is sufficiently close to one. On the contrary for δ close to 0 it can be positive as with the following numerical values for the model parameters:

$$n=4,\quad \gamma_1=\gamma_2=0.5,\quad \theta=0.2,\quad \delta=0.1,$$
$$\phi_2\approx -0.17,\quad \frac{\mathrm{d}\overline{G}_{-i}}{\mathrm{d}\overline{e}_i}\approx 0.57 .$$

[2] A linear FBNE is said to be convergent if, when the equilibrium decisions are replaced into the dynamical system, this system is globally asymptotically stable: whatever the initial conditions, the capital stocks converge to their steady state.

However a necessary condition for the dynamic model to tend to the static one is that δ be close to one, and in this case the conjectures are negative.[3]

Now let us try to explicit the relationship between these FBNE strategies and the conjectural variations equilibrium of the static contribution game. Any feedback Nash equilibrium strategy also satisfies Pontryagin's Maximum Principle, given that each individual considers the rival contributions be a function of the state variable G; therefore the following first order conditions are necessary:

$$\begin{aligned} U_c^i &= \mu_i \, , \quad \forall i \, , \\ \dot{\mu}_i &= (\theta + \delta)\, \mu_i - U_G^i - \mu_i \sum_{j \neq i} \frac{\mathrm{d}e_j}{\mathrm{d}G} \qquad \forall i \, . \end{aligned}$$

At the steady state the above conditions boil down to:

$$\frac{U_G^i}{U_c^i} = \theta + \delta - \frac{\mathrm{d}\overline{G}_{-i}}{\mathrm{d}\overline{G}} \, , \quad \forall i, \tag{2.13}$$

where $\frac{\mathrm{d}\overline{G}_{-i}}{\mathrm{d}\overline{G}} = \sum_{j \neq i} \frac{\mathrm{d}\overline{e}_j}{\mathrm{d}\overline{G}}$.

As $\theta \to 0$ and $\delta \to 1$ the above conditions can be rewritten:

$$\begin{aligned} \frac{U_c^i}{U_G^i} &= \frac{1}{1 - \frac{\mathrm{d}\overline{G}_{-i}}{\mathrm{d}\overline{G}}} \\ &= \frac{\mathrm{d}\overline{G}}{\mathrm{d}\overline{e}_i} = 1 + \frac{\mathrm{d}\overline{G}_{-i}}{\mathrm{d}\overline{e}_i} \, , \end{aligned}$$

for all i, which is corresponding to the conjectural equilibrium in the static game when $r = \frac{\mathrm{d}\overline{G}_{-i}}{\mathrm{d}\overline{e}_i}$. In this case the conjecture is not arbitrary anymore; it reflects a characteristic of the subgame perfect FBNE, *i.e.* that each individual takes into account how rivals react to a change in the state variable.

More explicitly, comparing (2.1) and (2.13), it follows:

$$1 + r \;=\; \frac{1}{1 - \frac{\mathrm{d}G_{-i}}{\mathrm{d}G}} \;=\; \frac{1}{1 - (n-1)\phi_2} \, .$$

This corresponds with the value previously found for r. The above identity highlights the idea that the conjectural parameter r captures all the underlying dynamic forces: the intertemporal trade-offs (as determined by

[3]Itaya and Shimomura also discuss the possibility of positive conjectures associated with non-linear feedback strategies.

the rate of time preference and the rate of depreciation) and the dynamic feedback interactions between individuals.

To summarise, the static Nash equilibrium turns out to coincide with the steady state of the dynamic game when θ tends to 0 and δ tends to 1, and contributors commit to path strategies. When on the contrary they play feedback Nash strategies, the steady state coincides with the outcome of a static conjectural variations game. The sign of the conjectures, still when θ and δ approach the values 0 and 1, is negative. This result is in sharp contrast with that of the trigger strategies in the repeated game model, which can easily account for positive conjectures.

2.3 Oligopoly

2.3.1 *Static Cournot oligopoly with constant conjectures*

There are n firms competing in the market of a homogeneous product. Denote by e_i the quantity produced by firm i and $E = \sum_j e_j$ the total quantity produced. The product price is related to industry output by an inverse demand function which we assume to be linear for simplicity:

$$p(E) = a - E, \quad a > 0 .$$

The firms' production technologies are summarised through a cost function $C_i(e_i)$.

Each firm holds some conjectures about how the rival quantity choices are affected by its own quantity decisions; more precisely, it is assumed that firm i conjectures that an infinitesimal variation of its quantity is to be followed by a change in total remaining quantities according to $\mathrm{d}E^c_{-i}/\mathrm{d}e_i = r$. As usual, $E_{-i} = \sum_{j \neq i} e_j$ stands for the production of the other firms.

Each firm's strategy results from the maximisation of its profit, given that it forms conjectures about rival reactions. Formally:

$$\max_{e_i} \; p\left(e_i + E^c_{-i}(e_i)\right) e_i - C_i\left(e_i\right) .$$

The optimising conditions yield:

$$p(E)\left[1 + (1+r)\,\frac{S_i}{\eta}\right] \;=\; C_i'(e_i)\;, \quad \forall i\;, \tag{2.14}$$

where $S_i = \frac{e_i}{E}$ represents firm i's market share, and $\eta = \frac{1}{p'}\frac{p}{E}$ is the price elasticity of demand. When $r = 0$ (Cournotian conjectures), one recognises the familiar mark-up pricing rule for the oligopoly. A non-zero conjecture modifies the importance of the elasticity of demand in the assessment of the marginal benefit. A positive value for the conjecture r means that if firm i contemplates the possibility to increment its quantities, this firm expects an increase of rival quantities and therefore a lower price; it is as if the price were more elastic to quantities and the incentives to produce are reduced. On the contrary a negative conjecture means less countervailing modifications of the rival strategies and higher incentives to produce.

In the particular case with linear costs (thus $C_i(e_i) = c_i e_i$, $\forall i$) and with linear inverse demand $p = a - E$, the above first order conditions can be rewritten as:

$$e_i = \frac{a - E_{-i} - c_i}{2 + r} \,, \qquad \forall i \,. \tag{2.15}$$

2.3.2 *A repeated linear oligopoly*

What could be the outcome(s) of a repeated version of this quantity game over an infinite horizon? In such a repeated game, firms aim at maximising their discounted stream of profits, with discount factor θ, and we assume away any conjectures about rivals' reactions. From the Folk theorem an infinity of Nash equilibria is known to exist. Cabral (1995) restricts his attention to a linear costs oligopoly with linear inverse demand $p = a - E$, and he focuses on the following class of trigger strategies: in each period, each firm sticks to some designated quantity $\overline{e}_i$; if in any period t a single firm i deviates, then this firm is punished by the other firms that select their quantities so as to drive firm i's profit down to zero. After one period of punishment, the firms return to the designated quantities. The optimality conditions to deter any single deviation e_i^* are:

$$\Pi^i \left(e_i^*, \overline{E}_{-i}\right) \;\leq\; (1+\theta)\, \Pi^i \left(\overline{e}_i, \overline{E}_{-i}\right) \,, \qquad i = 1, \ldots, n \,, \tag{2.16}$$

where $\overline{E}_{-i} = \sum_{j \neq i} \overline{e}_j$ and

$$e_i^* = \arg\max_{e_i} \Pi^i \left(e_i, \overline{E}_{-i}\right) .$$

Many equilibria exist in this class.[4] Among them are those with quantities $\overline{e}_i$ such that all the inequalities (2.16) are binding; then the optimality

[4] As noted before with the repeated contribution game, those strategies are not subgame perfect. Subgame perfection would require in particular that threats are themselves mutual best responses, which is not the case here.

conditions for the repeated oligopoly read as:

$$\Pi^i\left(e_i^*,\overline{E}_{-i}\right)=(1+\theta)\,\Pi^i\left(\overline{e}_i,\overline{E}_{-i}\right)\ ,\quad i=1,\ldots,n\ . \tag{2.17}$$

Computing $\Pi^i\left(e_i^*,\overline{E}_{-i}\right)$ one has:

$$\Pi^i\left(e_i^*,\overline{E}_{-i}\right)=\frac{1}{4}\left(a-\overline{E}_{-i}-c_i\right)^2\ ,$$

so that Condition (2.17) becomes

$$\begin{aligned}\frac{1}{4}\left(a-\overline{E}_{-i}-c_i\right)^2&=(1+\theta)\,\Pi^i\left(\overline{e}_i,\overline{E}_{-i}\right)\\&=(1+\theta)\left(a-\overline{E}_{-i}-\overline{e}_i-c_i\right)\overline{e}_i\ ,\end{aligned}$$

$i=1,\ldots,n$. Equivalently:

$$0\ =\ (1+\theta)\,\overline{e}_i^2-(1+\theta)\left(a-\overline{E}_{-i}-c_i\right)\overline{e}_i+\frac{1}{4}\left(a-\overline{E}_{-i}-c_i\right)^2\ .$$

Firm i's best response $\overline{e}_i\left(\overline{E}_{-i}\right)$ is a solution to this quadratic equation. There are two roots, and optimality requires to choose the smallest one, therefore:

$$\begin{aligned}\overline{e}_i&=\frac{(1+\theta)\left(a-\overline{E}_{-i}-c_i\right)}{2\,(1+\theta)}\\&\quad-\frac{\sqrt{(1+\theta)^2\left(a-\overline{E}_{-i}-c_i\right)^2-4\,(1+\theta)\,\frac{1}{4}\left(a-\overline{E}_{-i}-c_i\right)^2}}{2\,(1+\theta)}\\&=\frac{1}{2}\,\frac{a-\overline{E}_{-i}-c_i}{1+\theta+\sqrt{\theta(1+\theta)}}\ .\end{aligned}$$

Clearly from the last equation and Equation (2.15) the outcome of the repeated oligopoly corresponds exactly, and at each stage, to the conjectural variations equilibrium of the static game when $r=2(\theta+\sqrt{\theta(1+\theta)})$. In other words when the conjecture r takes on this particular value, the static conjectural oligopoly captures the more complex trigger strategies, involving one-period punishments, just described for infinitely repeated interactions.

In some cases these trigger strategies can support the cartel solution. Indeed, take the duopoly version of this game; the cartel solution emerges when $r=1$, or equivalently when $\theta=1/8$, *i.e.* the particular value that solves $1=2(\theta+\sqrt{\theta(1+\theta)})$. For some industries empirical supports exist for this cooperative prediction (see *e.g.* Slade (1995)).

2.3.3 Dynamic duopoly with adjustment costs

Driskill and McCafferty (1989) have extended the static symmetric duopoly analysis by adding adjustment costs. This is relevant when scaling up or down the output implies modifying the plant size; firms must then bear additional investment costs. Firm i's investment denoted x_i is thought to be the rate of change in output, therefore $\dot{e}_i = x_i$. Driskill and McCafferty (1989) have assumed quadratic adjustment costs:

$$A(x_i) = \frac{k}{2} x_i^2 \,, \qquad k > 0 \,.$$

When k tends to 0, the model boils down to the static duopoly analysed in the previous section, with the quantities e_i as the decision variables. However with strictly positive adjustment costs the duopoly is a linear-quadratic differential game, where the decision variables are no longer the e_is, but the x_is (that is, the rate of change of the former decision variables), and whose payoffs are:

$$\int_0^\infty e^{-\theta t} \left[p(e_i(t), e_j(t)) e_i(t) - C(e_i(t)) - A(x_i(t)) \right] \, dt \quad i = 1, 2 \,,$$

where $\theta > 0$ is the common discount rate, and with given initial conditions $e_i(0) = e_{i0}, i = 1, 2$.

Section 2.7.1 singles out necessary conditions for the existence of a linear convergent FBNE in a class of capital accumulation games that encompasses this one. Driskill and McCafferty (1989) show in addition that this equilibrium is unique. Let us denote it as follows:

$$x_i(t) = \phi_1 + \phi_2 \, e_i(t) + \phi_3 \, e_j(t) \qquad \phi_2, \phi_3 < 0 \,.$$

Notice that the equilibrium investment strategies are decreasing functions of the rival output, and yield stable trajectories for the quantities.

Dockner (1992) has analysed lengthily the links between the steady state of this dynamic duopoly and the outcome of its static counterpart. Let us follow Dockner and analyse in details the dynamic behaviours. Firm i's optimisation problem can be solved using Pontryagin's Maximum Principle, taking as given that the rival strategies are a linear feedback rule of the type $x_j(t) = x_j(e_i(t), e_j(t)) = \phi_1 + \phi_2 e_j(t) + \phi_3 \, e_i(t)$. Omitting the time argument, the current-value Hamiltonian is:

$$H^i = p(e_i, e_j) e_i - C(e_i) - A(x_i) + \lambda_i^i \, x_i + \lambda_j^i \, x_j(e_i, e_j) \,,$$

where the λ_j^i are the current-value adjoint variables. The necessary conditions are:

$$
\begin{aligned}
i)\ \frac{\partial H^i}{\partial x_i} &= 0 \Leftrightarrow A'(x_i) = \lambda_i^i\,, \\
ii)\ \dot{\lambda}_i^i &= \theta\,\lambda_i^i - p'e_i - p + C' - \lambda_j^i\,\frac{\partial x_j}{\partial e_i}\,, \\
&= \theta\,\lambda_i^i - p'e_i - p + C' - \lambda_j^i\,\phi_3\,, \\
iii)\ \dot{\lambda}_j^i &= \theta\,\lambda_j^i - p'e_i - \lambda_j^i\,\frac{\partial x_j}{\partial e_j}\,, \\
&= \theta\,\lambda_j^i - p'e_i - \lambda_j^i\,\phi_2\,.
\end{aligned}
$$

Let us analyse the steady state for which $\dot{e}_i = x_i = 0$. Then from $i)$ one can deduce $\lambda_i^i = 0$. The equation $\dot{\lambda}_j^i = 0$ produces:

$$\lambda_j^i = \frac{p'e_i}{\theta - \phi_2}\,.$$

Reporting this expression into the equation $\dot{\lambda}_i^i = 0$, and using the fact that $\lambda_i^i = 0$, one has:

$$p'e_i + p + \frac{p'e_i}{\theta - \phi_2}\,\phi_3 = C'\,,$$

or after some manipulations

$$p(E)\left[1 + \left(1 + \frac{\phi_3}{\theta - \phi_2}\right)\frac{S_i}{\eta}\right] = C'(e_i)\,.$$

This last expression is very similar to (2.14) which characterises the static duopoly CVE. Actually it is readily seen that the two outcomes are identical when the constant conjecture r in the static framework is equal to the quantity $\frac{\phi_3}{\theta-\phi_2}$, which is negative. To put it differently there exists a constant negative conjecture $r = \frac{\phi_3}{\theta-\phi_2}$ such that the feedback Nash equilibrium converges to a steady state which corresponds to the CVE of the associated static duopoly. As with the public good game, this finding is to be contrasted with the positive conjectures to be associated to the trigger strategies of the repeated oligopoly.

When firms commit to path strategies the adjoint equations (associated to the OLNE) are simpler

$$i)\ \frac{\partial H^i}{\partial x_i} = 0 \Leftrightarrow A'(x_i) = \lambda_i^i \ ,$$
$$ii)\ \dot{\lambda}_i^i = \theta\, \lambda_i^i - p' e_i - p + C' \ ,$$
$$iii)\ \dot{\lambda}_j^i = \theta \lambda_j^i - p' e_i \ .$$

A steady state OLNE is solution to $\dot{e}_i = x_i = \dot{\lambda}_i^i = \dot{\lambda}_j^i = 0$. Therefore *ii)* reduces to:

$$-p' e_i - p + C' = 0 \ ,$$

or equivalently

$$p\left(1 + \frac{S_i}{\eta}\right) = C' \ ,$$

i.e. the static mark-up rule of the Cournot-Nash solution. This equivalence holds for any values of the dynamic parameters k and θ. This is a surprising observation since, remember, in the public good example such an equivalence requires that some parameters of the dynamic structure take specific values. Actually the perfect equivalence in the above analysis is not robust; the reader can check that it no longer holds with linear-quadratic costs like $A(x_i) = \frac{k}{2} x_i^2 + l x_i,\ k, l > 0$.

2.4 Public infrastructure competition

As a last illustration one finds in Figuières (2002) a dynamic extension of a model of public capital competition proposed by Wildasin (1991).

2.4.1 *Static infrastructure competition*

The paper of Wildasin (1991) considers two jurisdictions denoted by $i = 1, 2$. The representative agent in jurisdiction i is endowed with an utility function defined over each possible pair of a consumption good c_i and an index s_i of environmental services:

$$u_i(c_i, s_i) = c_i + s_i \ .$$

The index of environmental services stems from the public infrastructures e_1 and e_2 built respectively by jurisdictions 1 and 2. In other words e_1 and

e_2 are the inputs in the joint production process of s_i

$$s_i = P^i(e_i, e_j) \ .$$

As a concrete illustration imagine two jurisdictions sharing the same waters (or the same international air-space) and building environmental infrastructures; the environmental policies undertaken to reduce pollution in jurisdiction i affect the welfare of jurisdiction j. We assume the following properties hold:

$A1)$ $P_i^i > 0$, the index of jurisdiction i is increasing with its own stock of infrastructure,

$A2)$ the environmental technology features decreasing returns to scale, *i.e.* the function $P^i(.,.)$ is increasing and concave with respect to e_i,

$A3)$ $P_j^i > 0$, the externalities between jurisdictions' infrastructures are positive.

Each jurisdiction also benefits from an exogenous revenue I_i through lump-sum taxes to finance the consumption of the private good and to buy the infrastructure at a price q:

$$c_i + qe_i = I_i \ . \tag{2.18}$$

The substitution of the budget constraint (2.18) in the utility function yields a payoff function:

$$V^i(e_i\,, e_j) = I_i + P^i(e_i\,, e_j) - qe_i \ .$$

and as usual, if agent i conjectures a relationship $e_j = \rho_j^c(e_i)$ between the decision variables, an interior CVE is solution to:

$$P_i^i(e_i\,, e_j) \ + \ r \ P_j^i(e_i\,, e_j) = q \ ,$$

where $r = (\rho_j^c)'$.

2.4.2 *Dynamic infrastructure competition*

In the dynamic version, at each point in time jurisdiction i can invest Q_i to change her public stock of infrastructures, so the law of motion for e_i is:

$$\dot{e}_i \ = \ Q_i \ .$$

Note again that e_i is no longer the decision variable; it is seen as a stock, and the decision variable of jurisdiction i is the rate of change Q_i of this stock. Investment entails an extra adjustment cost: in addition to the price to be paid for the infrastructure, the removal or modification of the existing stock cannot be done at no cost, workers must be hired for this task, and so on... This cost function is increasing and convex :

$$A(Q_i) = \frac{k}{2} Q_i^2 \,, \quad k \geq 0 \,,$$

and it is measured in the same terms as the utility of the consumption good and s_i. When the parameter k tends to 0 the game boils down to the static competition.

The flow budget constraints must now cover this additional cost:

$$c_i + A(Q_i) + qe_i \;=\; I_i \,. \tag{2.19}$$

The substitution of the budget constraint (2.19) in the utility function yields a reduced-form utility function:

$$V^i(e_i\,, e_j, Q_i) = I_i + P^i(e_i\,, e_j) - qe_i - A(Q_i)\,, \qquad i = 1, 2.$$

Intertemporal welfare in each jurisdiction is the discounted "sum" of the instantaneous criteria.

In order to keep the analysis tractable let us focus on a quadratic technology P^i such as:

$$P^i(e_i, e_j) = a(e_i + e_j) - (2\,e_i - e_j)^2, \quad a > 0, \quad i = 1, 2, \ j = 1, 2, \ i \neq j.$$

Under suitable restrictions on the parameter a the assumptions $A1$, $A2$ and $A3$ are satisfied. Clearly, a linear-quadratic differential game emerges, where the dynamics are:

$$\dot{e}_i = Q_i \,, \qquad e_i(0) = e_{i0} \,, \qquad i = 1, 2$$

and the payoffs are:

$$\int_0^\infty e^{-\theta\, t} \left[a\,(e_i(t) + e_j(t)) - (2\,e_i(t) - e_j(t))^2 - qe_i(t) - \frac{c}{2}\, Q_i^2 \right] \,\mathrm{d}t \,.$$

The constants I_i have been omitted without loss of generality.

Figuières (2000) provides some conditions to ensure the existence of a unique convergent linear FBNE of the type:

$$Q_i(t) = \phi_1 + \phi_2\, e_i(t) + \phi_3\, e_j(t), \qquad i = 1, 2, \quad j = 1, 2, \quad i \neq j \,.$$

It is also shown that the feedback coefficients ϕ_2 and ϕ_3 are respectively negative and positive. Remember that they were both negative in the context of duopoly models (see the next section for a general characterisation of the sign of ϕ_3).

Omitting the time argument, the current-value Hamiltonians associated to a FBNE are:

$$H^i = P^i(e_i\,,e_j) - qe_i - A(Q_i) + \lambda_i^i\, Q_i + \lambda_j^i\, Q_j\,(e_i\,,e_j)\ ,$$

where the λ_j^i are the current-value adjoint variables. The necessary conditions are:

$$
\begin{aligned}
i)\ \frac{\partial H^i}{\partial Q_i} = 0 \quad &\Leftrightarrow \quad A'(Q_i) = \lambda_i^i\ ,\\
ii)\ \dot{\lambda}_i^i &= \theta\, \lambda_i^i - P_i^i(e_i\,,e_j) + q - \lambda_j^i\, \frac{\partial Q_j}{\partial e_i}\ ,\\
&= \theta\, \lambda_i^i - P_i^i(e_i\,,e_j) + q - \lambda_j^i\, \phi_3\ ,\\
iii)\ \dot{\lambda}_j^i &= \theta \lambda_j^i - P_j^i(e_i\,,e_j) - \lambda_j^i\, \frac{\partial Q_j}{\partial e_j}\ ,\\
&= \theta \lambda_j^i - P_j^i(e_i\,,e_j) - \lambda_j^i\, \phi_2\ .
\end{aligned}
$$

At the steady state, $\dot{e}_i = \dot{Q}_i = \dot{\lambda}_i^i = \dot{\lambda}_j^i = 0$. From *i*) one can deduce $\lambda_i^i = 0$, therefore *ii*) simplifies to:

$$0 = -P_i^i(e_i\,,e_j) + q - \lambda_j^i\, \phi_3\ .$$

From $\dot{\lambda}_j^i = 0$ it follows that:

$$\lambda_j^i = \frac{P_j^i(e_i\,,e_j)}{\theta - \phi_2}\ .$$

Reporting this expression into the equation $\dot{\lambda}_i^i = 0$, and using the fact that $\lambda_i^i = 0$, one has:

$$P_1^i(e_i\,,e_j) + \frac{\phi_3}{\theta - \phi_2} P_2^i(e_i\,,e_j) = q\ ,$$

where the quantity $\frac{\phi_3}{\theta-\phi_2}$ is positive. When $r = \frac{\phi_3}{\theta-\phi_2}$, this last expression is identical to the one characterising the static conjectural game, with a positive (non-arbitrary) conjecture.

When firms commit to path strategies, the necessary conditions for an OLNE reduce to

$$i)\ \frac{\partial H^i}{\partial Q_i} = 0 \ \Leftrightarrow \quad A'(c_i) = \lambda_i^i\,,$$
$$ii)\ \dot{\lambda}_i^i = \theta\,\lambda_i^i - P_i^i(e_i\,,e_j) + q\,,$$
$$iii)\ \dot{\lambda}_j^i = \theta\lambda_j^i - P_j^i(e_i\,,e_j)\,.$$

A steady state OLNE is solution to $\dot{e}_i = Q_i = \dot{\lambda}_i^i = \dot{\lambda}_j^i = 0$. Therefore *ii*) reduces to:

$$P_i^i(e_i\,,e_j) = q\,,$$

which is exactly the first order condition for an interior Nash equilibrium in the static game.

Let us summarise those findings: *i*) feedback strategies are akin to strictly positive conjectures, *ii*) endogenous zero conjectures would capture instead commitment strategies in the truly dynamic setting.

2.5 A class of state-space games and the associated static games with conjectural variations

2.5.1 *A linear-quadratic framework with two state variables*

From the previous sections can we deduce that the CVE model is a useful shortcut for dynamic interactions? Can the modeller use it to predict whether, in a more realistic intertemporal framework, behaviours are more or less competitive? There is one obvious weakness in the argument followed so far, which the chapter illustrates: to know which conjecture is consistent with dynamic moods of play, one must first analyse the associated dynamic game, which is exactly what one would like to avoid! To overcome this problem, we ought to be able to deduce from the fundamentals of the problem at hand the nature of dynamic behaviours, and the associated sign of the conjecture. Figuières (2000) has studied a class of differential games with two-state variables that generalises Driskill and McCafferty (1989), Dockner (1992) and Figuières (2002), among others. For this whole class it is possible to associate static games, when some key parameters tend to zero, and then to figure out the relationship between its conjectural variations equilibrium and the value of the dynamic equilibria in the long run. This extension is interesting in that it allows one to identify which subclass of dynamic games is consistent with positive conjectures, and which one is consistent with negative conjectures.

The abstract class of games under consideration is as follows. Two players invest over an infinite horizon to build their stocks. Time is continuous and the states evolve according to:

$$\begin{cases} \dot{e}_i(t) = Q_i(t) - be_i(t), \\ e_i(0) = e_{0i}, \end{cases} \qquad i = 1, 2, \tag{2.20}$$

where $e_i(t) \in \mathbb{R}$ and $Q_i(t) \in \mathbb{R}$ represent, respectively, player i's capital stock and physical investment. The parameter $b \geq 0$ is the physical depreciation rate.

At each point of time, the two capital stocks contribute to player i's instantaneous payoffs according to a function $P^i(e_i, e_j)$, which we assume to be quadratic, and symmetric in the sense $P^j(e_j, e_i) = P^i(e_j, e_i)$. All the previous two-state variables frameworks we have seen so far are particular cases of the model presented here where the general expression for the function P^i is:

$$P^i(e_i, e_j) = a_0 + a_1\, e_i + a_2\, e_j + \frac{a_3}{2}\, e_i^2 + a_4\, e_i e_j + \frac{a_5}{2}\, e_j^2\,, \tag{2.21}$$

with $a_0 \geq 0, a_1 > 0,\ a_3 < 0,\ a_5 \leq 0$. The positive sign of a_1 ensures that, at least for some range of the capital stocks, the functions $P^i\,(.,.)$ increase with e_i. The other two assumptions on the parameters imply that the functions $P^i(.,.)$ are strictly concave with respect to player i's stock and also concave with respect to the rival capital stock. There are no restrictions *a priori* on the scalars a_2 and a_4, whose signs will depend on the context.

The costs of investment are:

$$C(Q_i) = \frac{c}{2}\, Q_i^2 + q\, Q_i\,, \qquad c > 0\,, \quad q \geq 0\,. \tag{2.22}$$

Each player associates an inter-temporal payoff to every vector of outcomes $(e_1(.), Q_1(.), e_2(.), Q_2(.))$:

$$\int_0^\infty e^{-\theta t}\left[P^i(e_i(t), e_j(t)) - C(Q_i(t))\right]\, \mathrm{d}t \tag{2.23}$$

for $i = 1, 2,\ j = 1, 2,\ i \neq j$. The parameter $\theta > 0$ stands for the common discount rate.

When c tends to 0 (but remains positive) and q tends to 0, the game is transformed into a static one, a "limit game" where the decision variables are the stocks e_i, and the payoffs are $P^i(e_i, e_j)$ (see also Driskill and McCafferty (1989) for a similar discussion). Computing the symmetric conjectural variations equilibrium of this static game, it follows that:

$$e^c = \frac{a_1 + a_2 r}{-(a_3 + a_4) - (a_4 + a_5) r}, \tag{2.24}$$

where $r \equiv (\rho_j^c)'$ is the constant conjecture.

When the parameter c is strictly positive, the costs of investment are increasing and convex ; intuitively one can expect that changes in the stocks will take place gradually over time.

As before we need to analyse the feedback Nash equilibrium. The feedback stationary strategy space for player i is the space of functions of the stocks:

$$\begin{aligned} Q_i : \mathbb{R} \times \mathbb{R} \to \quad & \mathbb{R}, \\ (e_i, e_j) \mapsto & Q_i(e_i, e_j) . \end{aligned}$$

Let us call Γ^f the class of feedback games that can be defined using (2.20)–(2.23) and the above feedback strategy space.

A feedback Nash equilibrium (Q_1^*, Q_2^*) can be characterised using a dynamic programming approach. The payoff function of player i is defined in (2.23), in which the law of motion for the state variables satisfies $\dot{e}_i = Q_i - b\, e_i$, $i = 1, 2$. The value functions $W^i(e_i, e_j)$, started at (e_i, e_j) when continuous and differentiable, and the FBNE solve the following HJB equations at each point of time :

$$\begin{aligned} \theta\, W^i(e_i, e_j) = \max_{Q_i} \{ & P^i(e_i, e_j) - C(Q_i) + W_i^i(e_i, e_j)(Q_i - b\, e_i) \\ & + W_j^i(e_i, e_j)(Q_j^*(e_i, e_j) - b\, e_j) \}, \ i = 1, 2 . \end{aligned} \tag{2.25}$$

Since the symmetric class of games under consideration is linear-quadratic, we shall look for a symmetric linear FBNE $(Q_1^*(e_1, e_2), Q_2^*(e_2, e_1))$. As previously explained, this FBNE will be convergent if, when the equilibrium decisions are plugged into the system (2.20), this system is globally asymptotically stable: whatever the initial conditions, the capital stocks converge to their steady state.

The first order conditions[5] for the maximisation of the HJB equations and the stability requirement mentioned above lead to:

[5]Those first order conditions are sufficient for maximisation since the expression to be maximised in (2.25) is concave in Q_i.

Theorem 2.1 *Take a game $\gamma^f \in \Gamma^f$ and let*

$$\begin{aligned}
Q_1^*(t) &= \frac{v^* - q}{c} + \frac{x^*}{c}\, e_1(t) + \frac{y^*}{c}\, e_2(t) \\
Q_2^*(t) &= \frac{v^* - q}{c} + \frac{x^*}{c}\, e_2(t) + \frac{y^*}{c}\, e_1(t) \\
W^i\left(e_i(t), e_j(t)\right) &= u^* + v^*\, e_i(t) + w^*\, e_j(t) + \frac{x^*}{2}\, e_i^2(t) \\
&\quad + y^*\, e_i(t)\, e_j(t) + \frac{z^*}{2}\, e_j^2(t)\ ,
\end{aligned}$$

where $u^*, v^*, w^*, x^*, y^*, z^*$ are solution to

$$\begin{aligned}
\theta\, u &= a_0 + \frac{v^2}{2c} + \frac{q^2}{2c} - \frac{q}{c}(v + w) + \frac{v\, w}{c} \\
\theta\, v &= a_1 + \frac{v\, x}{c} - q\frac{x + y}{c} - bv + \frac{v\, y}{c} + \frac{y\, w}{c} \\
\theta\, w &= a_2 + \frac{v\, y}{c} - q\frac{y + z}{c} + \frac{v\, z}{c} + \frac{x\, w}{c} - bw \\
\frac{\theta\, x}{2} &= \frac{a_3}{2} + \frac{x^2}{2c} - bx + \frac{y^2}{c} \\
\theta\, y &= a_4 + \frac{2\, x\, y}{c} + \frac{y\, z}{c} - 2by \\
\frac{\theta\, z}{2} &= \frac{a_5}{2} + \frac{y^2}{2c} + \frac{x\, z}{c} - bz \\
x &< bc \qquad (2.26) \\
y^2 &< (x - bc)^2\ . \qquad (2.27)
\end{aligned}$$

Then the pair $(Q_1^*(e_1, e_2),\, Q_2^*(e_2, e_1))$ is a convergent and symmetric linear FBNE for the game γ^f, and $W^i\left(e_i(t), e_j(t)\right)$ is player i's value function.

Proof: see Section 2.7.2.

One recognises in Theorem 2.1 the Riccati equations for the feedback coefficients of the value function. The sign of the feedback coefficient y^*/c conveys interesting information. Depending on the parameters it can be positive or negative. Intuitively, when it is negative it corresponds to situations where an increase of player i's stock reduces the player j's incentive to invest. On the other hand when y^*/c is positive an increase of e_i increases player j's incentive to invest. We shall come back in particular to the role of the coefficient y^* in the next section.

2.5.2 *Payoff structure and conjectures*

The static conjectural variations are meant to capture the effect of dynamic behaviours. Understanding those dynamic behaviours in a differential game is made difficult by the necessity to find out the FBNE, which necessitates to solve the Riccati equations, a system of non linear equations. Greater insights would be achieved if it were possible, like for static games, to foresee some of the qualitative properties of the equilibrium without having first to compute it, by the mere knowledge of the components of the game, and especially the knowledge of the payoff structure. In particular we would like to infer some information about the steady state. This section shows that this is possible for the class of games Γ^f.

Let us use the tools of Optimal Control Theory to determine the FBNE steady state. When agents use feedback strategies, player i's problem can be stated as follows. Given that player j's strategy is

$$Q_j^*(e_i, e_j) = \frac{v - q^*}{c} + \frac{x^*}{c} e_j + \frac{y^*}{c} e_i ,$$

player i aims at maximising (2.23) subject to (2.20). Introducing the costate variables μ_{ii} and μ_{ij}, player i's Hamiltonian is

$$\begin{aligned}\tilde{H}^i &= a_0 + a_1 e_i + a_2 e_j + \frac{a_3}{2} e_i^2 + a_4 e_i e_j + \frac{a_5}{2} e_j^2 \\ &\quad - \frac{c}{2} Q_i^2 - q Q_i + \mu_{ii} (Q_i - b e_i) + \mu_{ij} \left(Q_j^*(e_i, e_j) - b e_j \right) .\end{aligned}$$

The necessary conditions for optimality are:

$$\begin{cases} \dot{e}_i = \frac{1}{c} \mu_{ii} - b e_i - \frac{q}{c}, \qquad e_1(0) = e_{10} , \\ \dot{\mu}_{ii} = (\theta + b) \mu_{ii} - a_1 - a_3 e_i - a_4 e_j - \mu_{ij} \frac{\partial Q_j^*}{\partial e_i} , \\ \dot{\mu}_{ij} = (\theta + b - \frac{\partial Q_j^*}{\partial e_j}) \mu_{ij} - a_2 - a_4 e_i - a_5 e_j . \end{cases} \tag{2.28}$$

Given that the terms $\frac{\partial Q_j^*}{\partial e_i}$ and $\frac{\partial Q_j^*}{\partial e_j}$ are y^*/c and x^*/c respectively, the symmetric steady state $e_i = e_j = e^f$ of (2.28) can be calculated to be:

$$e^f = \frac{a_1 - (\theta + b) q + a_2 \frac{y^*/c}{\theta + b - x^*/c}}{(\theta + b) c b - (a_3 + a_4) - (a_4 + a_5) \frac{y^*/c}{\theta + b - x^*/c}} . \tag{2.29}$$

From the first stability condition (2.26) of the FBNE one can deduce

$$x^* - c(\theta + b) < x^* - cb < 0 \quad \Leftrightarrow \quad [x^* - c(\theta + b)]^2 > (x^* - cb)^2 \ .$$

From the above inequality and the second stability condition (2.27), it follows:

$$[x^* - c(\theta + b)]^2 > y^{*2}$$

$$\Leftrightarrow \ -1 < \frac{y^*}{x^* - c(\theta + b)} < 1 \quad \Leftrightarrow \quad -1 < \frac{y^*/c}{x^*/c - (\theta + b)} < 1 \ .$$

In addition, the sign of the quantity $\frac{y^*/c}{x^*/c-(\theta+b)}$ depends only on y^*. Indeed from the second stability condition

$$x^* - bc < 0 \ \Rightarrow \ (\theta + b)\, c - x^* > 0 \ .$$

It remains to determine the sign of y^*. This is the purpose of the following theorem:

Theorem 2.2 *Take a game $\gamma^f \in \Gamma^f$ and let*

$$Q_i^*(t) \ = \ \frac{v^* - q}{c} + \frac{x^*}{c}\, e_i(t) + \frac{y^*}{c}\, e_j(t) \ , \quad i = 1,2 \ j = 1,2 \ i \neq j \ ,$$

be a linear FBNE of γ^f. Then $sign(y^*) = sign(P_{ij}^i)$.

Proof: see Section 2.7.3.

In the class of games Γ^f the nature of feedback interactions at a FBNE can be deduced from the payoff structure in a straightforward way: when the function $P(e_i, e_j)$ is such that $P_{ij} < 0$ ($P_{ij} > 0$), the decision rules are decreasing (increasing) functions of the rival stock.

When the parameters θ and b, responsible for the time dependence, tend to zero the expression (2.29) of the steady state becomes:

$$e^f \ = \ \frac{a_1 - a_2(y^*/x^*)}{-(a_3 + a_4) + (a_4 + a_5)(y^*/x^*)} \ ,$$

which is formally the same as e^c given in (2.24) provided that

$$r \ = \ - \ \frac{y^*}{x^*} \ .$$

The same effect is obtained if q and c go to zero in (2.29). This value of the conjecture is not arbitrary, of course; we know that it is negative (positive) when $P_{ij} < 0$ ($P_{ij} > 0$).

From the above differential system it is also possible to characterise the open-loop Nash equilibrium. Indeed with an OLNE the term $\frac{\partial Q_j^*}{\partial e_i}$ is zero, and the OLNE steady state is:

$$e^o = \frac{a_1 - (\theta + b)q}{(\theta + b)cb - (a_3 + a_4)} \; .$$

When the dynamic structure is removed ($c \to 0$ and $q \to 0$), the OLNE steady state expression boils down to that of the static CVE with zero conjecture.

2.6 Conclusion

This chapter has presented different dynamic economic situations, the voluntary contributions to public goods, the oligopoly and the public infrastructures competition, which can be reinterpreted using the static conjectural variations model. What is learnt about reduced-form conjectures from those examples is as follows:

- positive conjectures capture more cooperative behaviours that are compatible with dynamic frameworks where players use trigger strategies. In some cases (infrastructure competition for instance) positive conjectures can also capture feedback behaviours. As shown in Section 2.5, this possibility occurs for well-identified payoffs structures (when $P^i_{ij} > 0$).
- Negative conjectures on the contrary capture the feedback behavioural characteristics in games where the payoff structure is such that $P^i_{ij} < 0$.
- Zero conjectures are akin to commitment strategies, at least for some key values of the dynamic parameters.

There exist many empirical evidences of collusive conjectures (see for instance Bordignon (1989) and Sugden (1982) in the context of the voluntary contribution to public goods, and Slade (1995) for oligopolies). To the extent that those conjectures capture dynamic behaviours, there is then a presumption in favour of underlying dynamic models that capture trigger strategies rather than feedback strategies, in particular when the payoff structure is consistent only with negative conjectures.

It would be interesting to appraise the influence of conjectures on long-term payoffs. This issue is challenged in Appendix B, where it appears that the sign of conjectures and externalities is a crucial information.

Finally it is worth stressing that the relationship between the static conjectural variations equilibria and the state-space games' outcomes holds only in the long run around the steady state. On the contrary the relationship with the repeated game equilibria holds at every point of time.

2.7 Technical complements

2.7.1 *The feedback Nash equilibrium in the voluntary contribution game*

We provide in this paragraph the proof of the assertions of Section 2.2.3. Let us consider a quadratic value function and check that it is indeed a solution to the Hamilton-Jacobi-Bellman equation (2.9). The value function takes the form:

$$W(G) = \phi_0 + \phi_1\gamma_1 G + \frac{\phi_2}{2}\gamma_1 G^2 \ , \tag{2.30}$$

where the $\phi_i, i = 0, 1, 2$ are unknown coefficients to be determined. Differentiating this function one has:

$$W'(G) = \phi_1\gamma_1 + \phi_2\gamma_1 G \ . \tag{2.31}$$

Plugging (2.30) and (2.31) into the HJB equation (2.9) yields:

$$\begin{aligned} 0 = -\theta &\left[\phi_0 + \phi_1\gamma_1 G + \frac{\phi_2}{2}\gamma_1 G^2\right] + \alpha + \beta_1 I - \frac{\gamma_1}{2} I^2 \\ &+ \frac{(-\beta_1 + \gamma_1 I)^2}{\gamma_1} + (-\beta_1 + \gamma_1 I)(\phi_1 + \phi_2 G) + \beta_2 G \\ &- \frac{\gamma_1}{2}\left\{\frac{(-\beta_1 + \gamma_1 I)^2}{\gamma_1^2} + \frac{2(-\beta_1 + \gamma_1 I)}{\gamma_1}(\phi_1 + \phi_2 G)\right. \\ &\qquad \left. + \left(\phi_1^2 + 2\phi_1\phi_2 G + \phi_2 G + \phi_2^2 G^2\right)\right\} \\ &- \frac{\gamma_2}{2} G^2 + (\phi_1\gamma_1 + \phi_2\gamma_1 G)\left[n\frac{(-\beta_1 + \gamma_1 I)}{\gamma_1} + n(\phi_1 + \phi_2 G) - \delta G\right] . \end{aligned}$$

Since this equation must hold for every possible G, the constant term, the G-term and the G^2-term on the right-hand side must be zero. As a result one has the so-called Riccati equations:

$$\gamma_1\left(\frac{2n-1}{2}\right)\phi_2^2-\gamma_1\left(\frac{\theta}{2}+\delta\right)\phi_2-\frac{\gamma_2}{2}=0\,,$$
$$-\theta\gamma_1\phi_1+\beta_2+(2n-1)\,\gamma_1\phi_1\phi_2+\phi_2 n\,(-\beta_1+\gamma_1 I)-\gamma_1\phi_1\theta=0\,.$$

Solving this system of equations:

$$\phi_2=\frac{1}{2n-1}\left(\frac{\theta}{2}+\delta\pm\sqrt{\left(\frac{\theta}{2}+\delta\right)^2+\frac{\gamma_2}{\gamma_1}(2n-1)}\right)\,,$$
$$\phi_1=\frac{\beta_2+\phi_2 n\,(-\beta_1+\gamma_1 I)}{(\theta+\delta)\gamma_1-(2n-1)\,\gamma_1\phi_2}\,.$$

It can be checked that only the negative root for ϕ_2 is relevant: only with this root is the resulting differential system globally asymptotically stable.

2.7.2 *Proof of Theorem 2.1*

In order to characterise the FBNE $(Q_1^*(e_1,e_2),Q_2^*(e_1,e_2))$ we will make use of dynamic programming in continuous time. The payoff of agent i is defined as follows:

$$\int_0^\infty e^{-rt}\,\left[P^i(e_i(t),e_j(t))-C(Q_i(e_i(t),e_j(t)))\right]\,\mathrm{d}t$$

where the state of the system is governed by $(Q_1(e_1,e_2),Q_2(e_1,e_2))$. The HJB equations give necessary and sufficient conditions to characterise the feedback Nash equilibrium:

$$\theta W^i(e_i,e_j)=\max_{Q_i}\{P^i(e_i,e_j)-C(Q_i)+W_i^i(e_i,e_j)(Q_i-b\,e_i)$$
$$+W_j^i(e_i,e_j)(Q_j^*-b\,e_j)\}\,,\qquad i=1,2\,.$$

The first order condition for the maximisation of the right-hand side of the HJB equation is:

$$-cQ_i^*-q+W_i^i(e_i,e_j)=0\quad\Leftrightarrow\quad Q_i^*=\frac{1}{c}W_i^i(e_i,e_j)-\frac{q}{c}\,.\qquad(2.32)$$

If we report (2.32) into the HJB equations, we obtain the following system with partial derivatives:

$$\theta W^i = P^i(e_i, e_j) + \frac{1}{2c}(W^i_i)^2 + \frac{q^2}{2c} - \frac{q}{c}(W^i_i + W^i_j) - bW^i_i e_i + \frac{1}{c} W^i_j W^j_j - bW^i_j e_j ,$$

for $i = 1, 2$. Because of the linear-quadratic nature of the problem one can "guess" quadratic value functions :

$$W^i(e_i, e_j) = u + v\, e_i + we_j + \frac{x}{2} e_i^2 + y\, e_i\, e_j + \frac{z}{2} e_j^2$$

for which:

$$\begin{aligned} W^i_i &= v + x\, e_i + y\, e_j , & W^i_j &= w + y\, e_i + z\, e_j , \\ W^j_j &= v + x\, e_j + y\, e_i , & W^j_i &= w + y\, e_j + z\, e_i . \end{aligned}$$

Reporting those expressions in the HJB equations, one obtains:

$$\begin{aligned} &\theta r\, u + \theta\, v\, e_i + \theta we_j + \theta\, \frac{x}{2}\, e_i^2 + \theta\, y\, e_i\, e_j + \theta\, \frac{z}{2}\, e_j^2 \\ &= a_0 + a_1 e_i + a_2 e_j + \frac{a_3}{2} e_i^2 + a_4 e_i e_j + \frac{a_5}{2} e_j^2 \\ &\quad + \frac{1}{2c}(v + xe_i + ye_j)^2 + \frac{q^2}{2c} - \frac{q}{c}(v + w + (x+y)e_i + (y+z)e_j) \\ &\quad - b(ve_i + xe_i^2 + ye_ie_j) + \frac{1}{c}(w + ye_i + ze_j)(v + xe_j + ye_i) \\ &\quad - b\,(we_j + ye_ie_j + ze_j^2) . \end{aligned}$$

Since this equation must be verified for all (e_i, e_j), by identification of the terms in $e_i, e_j, e_i^2, e_ie_j, e_j^2$ and of the constants, we find the so-called coupled Riccati equations for the coefficients in the value functions:

$$\begin{aligned} \theta\, u &= a_0 + \frac{v^2}{2c} + \frac{q^2}{2c} - \frac{q}{c}(v + w) + \frac{v\,w}{c} \\ \theta\, v &= a_1 + \frac{v\,x}{c} - q\frac{x+y}{c} - bv + \frac{v\,y}{c} + \frac{y\,w}{c} \\ \theta\, w &= a_2 + \frac{v\,y}{c} - q\frac{y+z}{c} + \frac{v\,z}{c} + \frac{x\,w}{c} - bw \end{aligned}$$

$$\frac{\theta\, x}{2} = \frac{a_3}{2} + \frac{x^2}{2c} - bx + \frac{y^2}{c}$$
$$\theta\, y = a_4 + \frac{2\, x\, y}{c} + \frac{y\, z}{c} - 2by$$
$$\frac{\theta\, z}{2} = \frac{a_5}{2} + \frac{y^2}{2c} + \frac{x\, z}{c} - bz$$

which satisfy the coefficients of the value function at the equilibrium. Those equations correspond to the system given in Theorem 2.1. From (2.32) the FBNE strategies are then:

$$Q_i^* \;=\; \frac{v^* - q^*}{c} + \frac{x^*}{c}\, e_i + \frac{y^*}{c}\, e_j \;.$$

Plugging these strategies into (2.20) one obtains the linear differential system driving the state variable. The conditions for stability of such a system are well known: the eigenvalues of the linear part must have a negative real part. The Routh-Hurwitz conditions (in this case: negative trace and positive determinant, see *e.g.* Horn and Johnson (1991)) provide the stability conditions:

$$x^* - bc < 0 \;,$$
$$(x^* - bc)^2 - y^{*2} > 0 \;,$$

for a convergent equilibrium. These conditions, rewritten as (2.26) and (2.27), in association with the Riccati equations above, characterise a convergent linear FBNE. This proves Theorem 2.1.

2.7.3 *Proof of Theorem 2.2*

We will make use of the equations in Theorem 2.1. The second group of three equations are independent of the others; they are non-linear with respect to x, y and z. Multiplying by c these equations, yields, after rearrangement:

$$0 = -\frac{r\theta c\, x}{2} + \frac{a_3 c}{2} + \frac{x^2}{2} - bcx + y^2$$
$$0 = -\theta\, c\, y + a_4\, c + 2\, x\, y + y\, z - 2bcy$$
$$0 = -\frac{\theta\, c\, z}{2} + \frac{a_5 c}{2} + \frac{y^2}{2} + x\, z - bcz \;.$$

Solving the last equation for z:

$$z = \frac{a_5 c + y^2}{\theta c - 2(x - bc)} \, .$$

Notice first that when $P_{ij} = a_4 = 0$, from the second equation $y^* = 0$ is a solution. Then x^* solves the quadratic equation

$$0 = -\frac{\theta \, c \, x}{2} + \frac{a_3 c}{2} + \frac{x^2}{2} - bcx \, ,$$

and z^* follows.

Otherwise when $a_4 \neq 0$, making use of the stability conditions and the fact that $a_5 \leq 0$

$$z = \frac{a_5 \, c + y^2}{\theta c + 2(bc - x)} \leq \frac{y^2}{\theta c + 2(bc - x)} < \frac{(bc - x)^2}{2(bc - x)} = \frac{bc - x}{2} \, .$$

From this inequality and the second equation

$$\frac{a_4 c}{y} = 2(bc - x) + \theta c - z > \frac{3}{2}(bc - x) + \theta c > 0 \, .$$

Therefore $\frac{a_4 c}{y} > 0$, meaning that y^* and $P^i_{ij} = a_4$ have the same sign.

Chapter 3

Consistent Conjectures in Dynamic Settings

3.1 Introduction

The purpose of this chapter is to examine extensions of the notion of conjectures and consistency to dynamic games. Cautious readers of the epistemic discussion presented at the beginning of Chapter 1 may feel a bit skeptical about some of the material presented here. To our best knowledge, there has been no epistemic attacks against those dynamic models; they may or may not be problematic. Still, a monograph on conjectural variations models could not ignore those contributions, if only for historical reasons. It is not our purpose to investigate epistemic issues, but the least we can do is to underline the type of rationality each model requires from agents, and discuss related technical difficulties.

We review four models of dynamic interactions which have in common the ideas that *i)* players form conjectures on how the other players react (or would react) to their actions, and *ii)* conjectures should be *consistent* with facts, in the sense that the evolution of the game should be the same as what was believed before implementing the decisions. In other words, conjectures of each player and actual best response reactions of the other players should coincide. In that sense, they are in the spirit of Bresnahan's definition of consistency for static games. The presence of a dynamic structure, with repeated interactions, and the observation of what rivals have actually played, gives more substance to the principle of consistency as the coincidence of prior assumptions and observed facts. However, differences appear in the way consistency is enforced. In a first group of models, consistency holds globally: the conjecture is constant over time, and the best response is computed solving an optimal control problem. Consistency exists if conjectures and responses coincide to some extent. This is the approach of Friedman (1968) (see also Friedman (1977), chapter 5), Laitner (1980) and Fershtman and Kamien (1985).

The second idea is to assume that players consider the game as a static conjectural variations game at each instant in time. Consistency in the sense of Bresnahan is then used. This is the approach of Başar *et al.* (1986).

We begin the first group of models with the work of Fershtman and Kamien (1985), where the notions of conjectural variation and conjectural equilibria are incorporated into the theory of differential games. This provides a new interpretation of open-loop and closed-loop equilibria in such games. These results are presented in Section 3.2.2.

A similar point of view is developed in Laitner (1980), in the setting of an infinite-horizon, discrete-time game. Both the instantaneous payoff and the conjectures of players are a function of the current state (of all players) *and* the next state (of that player). This model is described in Section 3.2.3.

We continue with the model of Friedman (1968), who was the first to describe a theoretical framework for defining *consistent* conjectures, in the case of dynamic oligopoly games in discrete-time and with an infinite horizon. Given beliefs a player forms about the dynamics of the states of her opponents, this player solves an optimal control problem and computes a stationary feedback. There is consistency if this feedback coincides with the beliefs of the other players. However, no result of existence and no example of such an equilibrium is described in the literature. In Section 3.2.4, we present Friedman's principle, and then, in Section 3.2.5, we develop the computations in the linear-quadratic setting. We finally apply the results to the dynamic Cournot duopoly. The principal concepts used in these three papers are summarised in Section 3.2.1.

The idea of using Bresnahan's static *consistent* conjectural variations equilibrium recursively at each step is developed in Başar *et al.* (1986). We discuss these results in Section 3.3.

We conclude the chapter in Section 3.4.

3.2 Conjectures for dynamic games, equilibria and consistency

We begin with a general description highlighting the common features of several works defining conjectures and some type of consistency in dynamic games. The description uses the formalism of discrete-time games. Although the literature is mostly limited to repeated games, without a proper state and a natural dynamics, all important concepts can be defined in general with dynamic games. The continuous-time counterpart will be the topic of Section 3.2.2. The principal contribution of this survey is a new terminology in order to distinguish the different variants of conjectural equilibria used in the literature.

We then provide some details about the results obtained by Fershtman and Kamien (1985) (Section 3.2.2), Laitner (1980) (Section 3.2.3) and on the model of Friedman (1968) (Section 3.2.4).

3.2.1 *Principle*

Consider a dynamic game with n players and time horizon T, finite or infinite. The state of the game at time t is described by a vector $\mathbf{x}(t) = (x_1(t), \ldots, x_m(t)) \in \mathbb{R}^m$. Player i has an influence on the evolution of the state, through a control variable. Let $e_i(t)$ be the control performed by player i in the t-th period of the game, that is, between time t and $t+1$. Let $\mathbf{e}(t) = (e_1(t), \ldots, e_n(t)) \in \mathbb{E}$ denote the vector of controls applied by each player. The state of the game at time $t+1$ results from the combination of the controls $\mathbf{e}(t)$ and the state $\mathbf{x}(t)$. Formally, the state evolves according to some dynamics

$$\mathbf{x}(t+1) \;=\; f(\mathbf{x}(t), \mathbf{e}(t)) \,, \tag{3.1}$$

with an initial state $\mathbf{x}(0) = \mathbf{x}_0$. When the model is a repeated game, this setting can be used with the convention that the "state" coincides with the previous control, according to: $\mathbf{x}(t+1) = \mathbf{e}(t)$. The instantaneous payoff of player i is a function Π^i of the state and controls, and her total payoff is given by:

$$V^i(\mathbf{x}_0; \mathbf{e}(0), \mathbf{e}(1), \mathbf{e}(2), \ldots, \mathbf{e}(T-1)) \;=\; \sum_{t=1}^{T} \theta_i^{t-1}\, \Pi^i(\mathbf{x}(t), \mathbf{e}(t)) \,, \tag{3.2}$$

θ_i being a discount parameter.

Each player has a belief on the behaviour of the other ones. More precisely player i *thinks* that player j chooses her control by applying some function ϕ_t^{ij} to observed quantities. Several degrees of behavioural complexity are possible here. One may assume, as in Friedman (1968), that the conjecture involves only the last observed state, that is:

$$e_j^c(t) \;=\; \phi_t^{ij}(\mathbf{x}(t)) \,. \tag{3.3}$$

The superscript "c" attached to the next action $e_j(t)$ stands for "conjectured". One may also assume that the conjecture also involves the last actions of players. This writes as:

$$e_j^c(t) = \phi_t^{ij}(\mathbf{x}(t), \mathbf{e}_{-j}(t-1)) , \tag{3.4}$$

with $\mathbf{e}_{-i} = (e_1, \ldots, e_{i-1}, e_{i+1}, \ldots, e_n)$. This form means that player i believes that player j uses as information the current state and the last action of all players except, of course, player j herself. Another possible form, in the spirit of conjectural variations, is

$$e_j^c(t) = \phi_t^{ij}(\mathbf{x}(t), \mathbf{e}_{-j}(t)) . \tag{3.5}$$

This type of conjecture is called "complete" by Fershtman and Kamien (1985), who use it in a continuous-time model. In discrete-time, this is the form proposed by Laitner (1980). Here, the conjecture is that the other players somehow react to the control player i is about to choose.

Given this belief, player i concludes by replacing $e_j(t)$, for $j \neq i$, by $\phi_t^{ij}(\mathbf{x}(t))$ in the dynamics (3.1), that the state actually evolves according to some dynamics[1]

$$\mathbf{x}(t+1) = \tilde{f}^i(\mathbf{x}(t), e_i(t)) , \tag{3.6}$$

where only her own decision has an influence. Her maximisation problem is therefore a standard dynamic control problem. Solving it, player i obtains an optimal action profile $\{e_i^{i*}(t)\}$, assumed unique.[2] After replacing in her conjecture scheme ϕ_t^{ij}, she deduces conjectured actions of her opponents $\{e_j^{i*}(t)\}$, $j \neq i$, and a conjectured state path $\{\mathbf{x}^{i*}(t)\}$.

If player j actually implements the decision rule e_j^{j*}, the evolution of the state will follow the real dynamics (3.1), and result in some actual trajectory $\{\mathbf{x}^a(t)\}$. Players will observe a discrepancy with their beliefs unless the actual path coincides with their conjectured path. If it does, no player will have a reason to challenge their conjectures and deviate from the "optimal" control they have computed. This leads to the following definition of a *state-consistent* equilibrium. Denote by $\boldsymbol{\phi}_t^i$ the vector of functions $(\phi_t^{i1}, \ldots, \phi_t^{i,i-1}, \phi_t^{i,i+1}, \ldots, \phi_t^{in})$.

[1] Dynamics of the type $\mathbf{x}(t+1) = \tilde{f}^i(\mathbf{x}(t), e_i(t), \mathbf{e}(t-1))$ may result if conjectures are of the form (3.4). For conjectures of the type (3.5), the elimination of the $n-1$ variables $e_j(t)$ involves the resolution of a system of $n-1$ equations with parameters $e_i(t)$ and $\mathbf{x}(t)$. The reasoning requires that the solution be unique.

[2] We shall use the superscript "$i*$" as a shorthand for "believed by player i".

Definition 3.1 (STATE-CONSISTENT CONJECTURAL EQUILIBRIUM) The vector of conjectures $(\phi_t^1, \ldots, \phi_t^n)$ is a state-consistent conjectural equilibrium if

$$\mathbf{x}^{i*}(t) = \mathbf{x}^a(t) , \tag{3.7}$$

for all i and t, and for all initial state $\mathbf{x}(0) = \mathbf{x}_0$.

An alternative definition, proposed by Fershtman and Kamien (1985), requires the coincidence of control paths, given the initial condition of the state:

Definition 3.2 (WEAK CONTROL-CONSISTENT CONJECTURAL EQUILIBRIUM) The vector of conjectures $(\phi_t^1, \ldots, \phi_t^n)$ is a weak control-consistent conjectural equilibrium if

$$\mathbf{e}^{i*}(t) = \mathbf{e}^{j*}(t) , \tag{3.8}$$

for all $i \neq j$ and t, given the initial state $\mathbf{x}(0) = \mathbf{x}_0$.

The stronger notion proposed by Fershtman and Kamien (1985) (where it is termed the "perfect" equilibrium) requires the coincidence for all possible initial states:

Definition 3.3 (CONTROL-CONSISTENT CONJECTURAL EQUILIBRIUM) The vector of conjectures $(\phi_t^1, \ldots, \phi_t^n)$ is a control-consistent conjectural equilibrium if the coincidence of controls (3.8) holds for all $i \neq j$, all t and all initial state $\mathbf{x}(0) = \mathbf{x}_0$.

Clearly, any control-consistent conjectural equilibrium is state-consistent provided the dynamics (3.1) have a unique solution. It is of course possible to define a weak state-consistent equilibrium, where coincidence of trajectories is required only for some particular initial value. This concept does not seem to be used in the existing literature.

Another approach uses the fact that the solution of deterministic control problems can usually be expressed as a *state feedback*. Accordingly, when solving her (conjectured) optimisation problem, player i concludes that there exists a function ψ_t^i of the state such that:

$$e_i^{i*}(t) = \psi_t^i(\mathbf{x}(t)) .$$

Consistency can then be defined as the requirement that optimal feedbacks coincide with conjectures.

Definition 3.4 (FEEDBACK-CONSISTENT CONJECTURAL EQUILIBRIUM) The vector of conjectures $(\phi_t^1, \ldots, \phi_t^n)$ is a feedback-consistent conjectural equilibrium if $\psi_t^i = \phi_t^{ji}$ for all $i \neq j$ and all t.

Obviously, consistency in this sense implies that the conjectures of two different players about some third player i coincide (see also Section 1.3.9):

$$\phi_t^{ji} = \phi_t^{ki}, \quad \forall i \neq j \neq k, \quad \forall t. \tag{3.9}$$

In addition, if the time horizon T is infinite, and if there exists a *stationary* feedback ψ_∞^i, then a conjecture which is consistent with this stationary feedback should coincide with it at any time. This implies that the conjecture does not vary over time. For a simple equilibrium in the sense of Definition 3.1, none of these requirements are necessary *a priori*. It may happen that trajectories coincide in a "casual" way, resulting from discrepant conjectures of the different players.

3.2.2 *Fershtman and Kamien: conjectures in differential games*

The model of Fershtman and Kamien (1985) is a continuous-time, finite-horizon game. With respect to the general framework of the previous section, the equation of the dynamics (3.1) becomes

$$\dot{\mathbf{x}}(t) = f(\mathbf{x}(t), \mathbf{e}(t)),$$

and the total payoff is:

$$\int_0^T e^{-\theta_i t}\, \Pi^i(\mathbf{x}(t), \mathbf{e}(t))\, dt. \tag{3.10}$$

Players have conjectures of the form (3.3), or "complete conjectures" of the form (3.5). In continuous time, the conjectured value is that of the opponent's state at the current time:

$$e_j^c(t) = \phi_t^{ij}(\mathbf{x}(t), \mathbf{e}_{-j}(t)).$$

Conjectures are assumed to be the same for all players (Condition (3.9)). Classically, the definition of a dynamic game must specify the space of strategies with which players can construct their action $e_i(t)$ at time t. The information potentially available being the initial state $\mathbf{x}_0$ and the current state $\mathbf{x}(t)$, three classes of strategies are defined: *i*) closed-loop no-memory strategies, where $e_i(t) = \psi^i(\mathbf{x}_0, \mathbf{x}, t)$, *ii*) feedback Nash strategies where $e_i(t) = \psi^i(\mathbf{x}, t)$, and *iii*) open-loop strategies where $e_i(t) = \psi^i(\mathbf{x}_0, t)$.

The first concept of conjectural equilibrium studied is the one of Definition 3.2 (weak control-consistent equilibrium). The following results are then obtained:

- Open-loop Nash equilibria are weak control-consistent conjectural equilibria.
- Weak control-consistent conjectural equilibria are closed-loop no-memory equilibria.

In other words, the class of weak control-consistent conjectural equilibria is situated between open-loop and closed-loop no-memory equilibria.

Fershtman and Kamien further define *perfect* conjectural equilibria as in Definition 3.3: control-consistent equilibria. The result is then:

- Control-consistent conjectural equilibria and feedback Nash equilibria coincide.

Further results of the paper include the statement of the problem of calculating complete conjectural equilibria (defined as Definition 3.2 with conjectures of the form (3.5)). The particular case of a duopoly market is studied. The price is the state variable $x(t)$ of this model; it evolves according to a differential equation depending on the quantities $(e_1(t), e_2(t))$ produced by both firms. The complete conjectures have here the form: $\phi^i(x; e_j)$. The feedback Nash equilibrium is computed, as well as the complete conjectural equilibrium with affine conjectures. The two equilibria coincide when conjectures are actually constant. When $\partial\phi^i(x; e_j)/\partial e_j = 1$, the stationary price is the monopoly price.

Laitner (1980) analyses the same type of conjectures in discrete-time. We describe his approach in the following section.

3.2.3 *Laitner's discrete-time model with complete conjectures*

The model of Laitner (1980) is a discrete-time, infinite-horizon repeated game, with two players. Players are endowed with conjectures of the "complete" form (3.4). The point of view of players is therefore that other players will observe their play at the current stage, and base their reactions on it (as well as on the previous state of the game). Since all players are assumed to possess this behaviour, we have therefore the same "double Stackelberg" situation which is characteristic of the static conjectural equilibrium.

Laitner considers a payoff structure of the form:

$$V^i(\mathbf{x}_0; \mathbf{e}(0), \mathbf{e}(1), \ldots) = \sum_{t=0}^{\infty} \theta_i^t \, \Pi^i(\mathbf{x}(t), e_i(t)) \, , \tag{3.11}$$

which takes into account inter-temporal adjustment costs. Observe, with respect to our general presentation of Section 3.2.1, that since we are in a repeated game, the "state" $\mathbf{x}(t+1)$ coincides with the control $\mathbf{e}(t)$.

Player i has a conjecture on the value of $x_j(t+1)$, of the form: $x_j(t+1) = \phi^i(x_i(t), x_j(t); e_i(t))$.

Laitner's definition of an equilibrium is Definition 3.1 (state-consistency), taking into account that states and controls coincide since this is a repeated game with conjectures. Observe that, in that case, Definitions 3.1 and 3.3 actually coincide.

The determination of the optimal control of players is performed using the Hamilton-Jacobi-Bellman equations. Computations are carried out with a duopoly model in quantities, leading to the following profit form: for firm i,

$$\sum_{t=0}^{\infty} \theta^t \, (x_i(t) \;\; (\alpha + \beta(x_1(t) + x_2(t))) \;\; - \;\; \sigma(x_i(t)^2 + (x_i(t) - e_i(t))^2)) \; .$$

In addition, conjectures are assumed to be affine, of the form:

$$x_j = \phi^i(x_i, x_j; e) = a^i x_i + b^i x_j + c^i + d^i e \; .$$

Laitner proves the existence of an *infinite number* of distinct such equilibria. There is therefore an equilibrium selection problem. We come back to this issue in the conclusion of the chapter. Observe also that the situation in discrete time differs from the situation in continuous time. In continuous time, conjectures, plays and the observation (that is, the verification) of the outcome are simultaneous. In discrete-time, there is a lag of one time step. It may be that there is more room in discrete-time for a "casual" coincidence of conjectures and facts.

3.2.4 *Friedman's dynamically consistent conjectures*

The model of Friedman (1968) is a discrete-time, infinite-horizon repeated game with n players. The total payoff of player i has the form:

$$V^i(\mathbf{x}_0; \mathbf{e}(0), \mathbf{e}(1), \dots) = \sum_{t=1}^{\infty} \theta_i^{t-1} \, \Pi^i(\mathbf{x}(t)) \,, \qquad \mathbf{x}(t+1) = \mathbf{e}(t) \,,$$

for some discount factor θ_i. Players have conjectures of the form $x_j(t+1) = \phi^i(x(t))$. The equilibrium concept advocated (called a "reaction function equilibrium") is what we have defined as the *feedback-consistent equilibrium* in Definition 3.4.

In order to identify such equilibria, Friedman suggests to solve the control problem with a finite horizon T, and then let T tend to infinity to obtain a stationary optimal feedback control. If the finite-horizon solutions converge, this has the effect of selecting certain solutions among the possible solutions of the infinite-horizon problem. Once the stationary feedbacks ψ^i are computed, the problem is to match them with the conjectures ϕ^i.

The results obtained on this concept of equilibrium, in the context of oligopoly theory, are collected in Friedman (1977), Chapter 5. Conditions under which the optimal finite-horizon feedback controls do converge to an optimal stationary feedback have been established. The existence of a weaker form of feedback equilibria has also been proved; these equilibria have the property that there exists a strategy profile $\mathbf{x}$ such that, when evaluated at $\mathbf{x}$, the functions ψ^i and ϕ^i coincide, as well as their partial derivatives. This equilibrium is therefore similar to the CCVE of order 0 for static games (Definition 1.7).

However, no results have been reported so far about the existence (or non-existence) of feedback-consistent equilibria, except for the obvious one consisting in the repetition of the Nash equilibrium of the static game. Indeed, we have:

Theorem 3.1 (THE REPEATED STATIC NASH EQUILIBRIUM IS A FEEDBACK-CONSISTENT EQUILIBRIUM) *Assume there exists a unique Nash equilibrium $(e_1^N, \dots, e_2^N)$ for the one-stage (static) game. Then if some player i conjectures that the other players will play the strategies e_{-i}^N at each stage, then her own optimal response is unique and is to play e_i^N at each stage.*

Proof. Let (e_i^N, e_{-i}^N) denote the unique Nash equilibrium of the static game. Since player i assumes that her opponents systematically play e_{-i}^N, we have $x_{-i}(t) = e^N_{-i}$ for all t. Therefore, her perceived optimisation problem is:

$$\max_{\{e_i(0),e_i(1),\dots\}} \sum_{t=1}^{\infty} \theta_i^{t-1}\, \Pi^i\left(e_i(t-1), e^N_{-i}\right) \ .$$

Since e_i^N is the best response to e^N_{-i}, the optimal control of player i is $e_i(t) = e_i^N$ for all $t \geq 0$. In other words, player i should respond to her "Nash conjecture" by playing Nash repeatedly. □

3.2.5 *Feedback-consistency for linear-quadratic games*

In this section, we perform the calculations following Friedman's method in the case of linear-quadratic games with an arbitrary number of players.

Our principal motivation for doing this lies in the observation that the model developed by Laitner (1980) exhibits two major problems. First, the behaviour model of players has the "double Stackelberg" characteristic of static games: each player thinks that the other will react to a play she is about to make. Our point of view is that the analysis of dynamic models is necessary precisely to eliminate this type of dubious rationality. Second, Laitner's analysis concludes that there exists a continuum of equilibria to his model. The model of Friedman does not suffer from the first drawback: the decisions of players are based on observed values. However, no existence results for this model being known yet, it is unknown whether the model also has a multiplicity of equilibria. Since our computations conclude negatively for duopoly models, we demonstrate that the multiplicity of equilibria observed by Laitner, also in a duopoly model, depends closely on the way the agents' conjectures are modelled. On the other hand, we also exhibit the case of a convex distance game, where a multiplicity of feedback-consistent equilibria does exist.

The analysis will be done in several steps. First, we consider a finite time horizon game with conjectures, and we construct the optimal control of each player (Lemma 3.2). Next, we consider conjectures of a particular "proportional" form and we derive the conditions for the existence of a stationary feedback control (Theorem 3.5). The multiplicative coefficients of this control are always *proportional* to those of the conjecture (Lemma 3.3). We therefore obtain conditions for the existence of a consistent equilibrium in the sense of Definition 3.4, in the case of symmetric players (Theorem 3.7). Those conditions bear on the sign of quantities involving the coefficients of the conjectures and on the parameters of the model.

We next apply these results to two models: the duopoly of Cournot with constant marginal costs (Paragraph 3.2.5.5) and a distance game (Paragraph 3.2.5.6). The instantaneous payoff function is concave for this second model, but not for the first. For the duopoly, we prove that *there exist no other consistent equilibrium than the repeated static Nash equilibrium.* For the distance game however, we find a continuum of feedback-consistent equilibria, in addition to the repeated static Nash strategy.

3.2.5.1 *Setting of the problem*

Consider the discrete-time dynamic game with n players as in Section 3.2.1, where the instantaneous payoff of player i is an affine-quadratic function of the state. In matrix/vector notation, we have:

$$\Pi^i(\mathbf{x}) \;=\; \frac{1}{2}\,\mathbf{x}^t\mathbf{K}^i\mathbf{x} \;+\; \mathbf{L}^i\mathbf{x} \;+\; \mathbf{M}^i\;. \tag{3.12}$$

Here, $\mathbf{K}^i, \mathbf{L}^i, \mathbf{M}^i$ are $(n\times n)$, $(1\times n)$ and (1×1)-matrices, respectively. The state vector $\mathbf{x}$ is a $(n \times 1)$ column vector. We will denote in general the transposed vector of any vector $\mathbf{x}$ as $\mathbf{x}^t$. The matrices $\mathbf{K}^i$ are assumed to be symmetric but not necessarily positive. The total payoff of player i is

$$V^i(\mathbf{x}_0;\mathbf{e}(0),\mathbf{e}(1),\dots,\mathbf{e}(T-1)) \;=\; \sum_{t=1}^{T} \theta_i^{t-1}\;\Pi^i(\mathbf{x}(t))\;, \tag{3.13}$$

where θ_i is some discount parameter. At each time step t, player i controls freely the value of $e_i(t)$, and the next state has the value $x_i(t+1) = e_i(t)$. Player i thinks that player j will, at time $t+1$, choose her control (or her next state) according to the *affine* rule:

$$x_j(t+1) \;=\; e^c_j(t) \;=\; \sum_{k=1}^{n} f^i_{j,k}(\tau)\; x_k(t) \;+\; g^i_j(\tau)\;,$$

where $\tau = T-(t+1)$ denotes the number of time units left before the end of the game (starting from zero), and $t = 0,\dots,T-1$. The simultaneous use of the variables t and τ in this rule is a notational convenience. As we shall see, due to the "backwards" nature of the following construction, it is more convenient to measure time with τ.

The set of numbers $f^i_{j,k}$ and g^i_j, $1\le j\le n$, $i\neq j$, $1\le k\le n$ represents the conjecture that player i makes about the behaviour of other players.

Define the matrix $\mathbf{F}^i(\tau)$ and the vector $\mathbf{g}^i(\tau)$ with coordinates:

$$\mathbf{F}^i_{j,k}(\tau) = \begin{cases} f^i_{j,k}(\tau) & \text{if } i \neq j \\ 0 & \text{if } i = j \end{cases} \quad \text{and} \quad \mathbf{g}^i_j(\tau) = \begin{cases} g^i_j(\tau) & \text{if } i \neq j \\ 0 & \text{if } i = j \end{cases}$$

As we have seen, from the point of view of player i, the optimisation problem is: compute

$$W_T(\mathbf{x}_0) = \max_{\{e(0),\dots,e(T-1)\}} \{V^i(\mathbf{x}_0; \mathbf{x}(1), \mathbf{x}(2), \dots, \mathbf{x}(T))\} \tag{3.14}$$

under conditions $\mathbf{x}(t+1) = e(t)\mathbf{b}_i + \mathbf{F}^i(\tau)\mathbf{x}(t) + \mathbf{g}^i(\tau)$, $0 \leq t \leq T-1$ (with $\tau = T-(t+1)$) and $\mathbf{x}(0) = \mathbf{x}_0$. The function $W_T(\cdot)$ is the value function of the control problem. We have denoted here by $\mathbf{b}_i$ the column vector $(0, \dots, 1, \dots, 0)^t$ with the "1" in position i.

Along the analysis, we shall implicitly adopt the point of view of player 1, and drop the superscript "i". This will not necessarily mean that players are symmetrical.

3.2.5.2 *Optimal reaction*

Consider the matrices constructed with the following recurrence:

$$\mathbf{K}(0) = \mathbf{0}, \quad \mathbf{L}(0) = \mathbf{0}, \quad \mathbf{M}(0) = 0, \tag{3.15}$$

$$\mathbf{A}(\tau) = \mathbf{K} + \theta\mathbf{K}(\tau), \quad \mathbf{B}(\tau) = \mathbf{L} + \theta\mathbf{L}(\tau), \quad \mathbf{C}(\tau) = \mathbf{M} + \theta\mathbf{M}(\tau) \tag{3.16}$$

$$\xi(\tau) = \mathbf{b}_1^t\mathbf{A}(\tau)\mathbf{b}_1 = (\mathbf{A}(\tau))_{1,1} \tag{3.17}$$

$$\Phi(\tau) = \left(\mathbf{I} - \frac{1}{\xi(\tau)}\,\mathbf{b}_1\mathbf{b}_1^t\mathbf{A}(\tau)\right)\,\mathbf{F}(\tau) \tag{3.18}$$

$$\Psi(\tau) = \left(\mathbf{I} - \frac{1}{\xi(\tau)}\,\mathbf{b}_1\mathbf{b}_1^t\mathbf{A}(\tau)\right)\,\mathbf{g}(\tau) - \frac{\mathbf{B}(\tau)\mathbf{b}_1}{\xi(\tau)}\,\mathbf{b}_1 \tag{3.19}$$

$$\mathbf{K}(\tau+1) = \Phi^t(\tau)\mathbf{A}(\tau)\Phi(\tau) \tag{3.20}$$

$$\mathbf{L}(\tau+1) = \Psi^t(\tau)\mathbf{A}(\tau)\Phi(\tau) + \mathbf{B}(\tau)\Phi(\tau) \tag{3.21}$$

$$\mathbf{M}(\tau+1) = \frac{1}{2}\Psi^t(\tau)\mathbf{A}(\tau)\Psi(\tau) + \mathbf{B}(\tau)\Psi(\tau) + \mathbf{C}(\tau)\,. \tag{3.22}$$

This recurrence is well defined only if $\xi(\tau) \neq 0$ for all $\tau = 0, \dots, T-1$. Observe that matrices $\mathbf{A}(\tau)$ and $\mathbf{K}(\tau)$ are symmetric.

Lemma 3.2 *If $\xi(\tau) < 0$ for all $\tau = 0, \dots, T-1$, then the solution of player 1's optimisation problem is given by the sequence of states:*

$$\mathbf{x}(t+1) = \Phi(\tau)\ \mathbf{x}(t)\ +\ \Psi(\tau)\ , \tag{3.23}$$

$t = 0, \dots, T-1$, $\tau = T-(t+1)$, *with in particular the optimal control*

$$e(t)\ =\ \mathbf{b}_1^t\,(\Phi(\tau)\ \mathbf{x}(t)\ +\ \Psi(\tau))\ .$$

The value function is given by:

$$W_T(\mathbf{x})\ =\ \frac{1}{2}\ \mathbf{x}^t\mathbf{K}(T)\mathbf{x}\ +\ \mathbf{L}(T)\mathbf{x}\ +\ \mathbf{M}(T)\ . \tag{3.24}$$

If $\xi(\tau) > 0$ for some τ, there does not exist an optimal control, and $W_T(\mathbf{x}) = +\infty$.

Proof. The proof proceeds by recurrence simultaneously for Equations (3.24) and (3.23). Obviously, for $T = 0$, the payoff (3.13) of player 1 is 0, so that (3.24) holds with the matrices defined in (3.15).

Assume now that (3.24) holds for some value of τ. According to the dynamic programming principle, we have:

$$W_{\tau+1}(\mathbf{x}) = \max_e\{\Pi(\mathbf{q})\ +\ \theta W_\tau(\mathbf{q})\} \tag{3.25}$$

$$\mathbf{q} = e\ \mathbf{b}_1\ +\ \mathbf{F}(\tau)\ \mathbf{x}\ +\ \mathbf{g}(\tau)\ . \tag{3.26}$$

Using (3.24) and definitions (3.16), this becomes:

$$\begin{aligned} W_{\tau+1}(\mathbf{x}) &= \max_e\left\{\frac{1}{2}\ \mathbf{q}^t\mathbf{K}\mathbf{q}\ +\ \mathbf{L}\mathbf{q}\ +\ \mathbf{M}\right. \\ &\qquad \left.+\ \theta\left(\frac{1}{2}\ \mathbf{q}^t\mathbf{K}(\tau)\mathbf{q}\ +\ \mathbf{L}(\tau)\mathbf{q}\ +\ \mathbf{M}(\tau)\right)\right\} \\ &= \max_e\left\{\frac{1}{2}\ \mathbf{q}^t\mathbf{A}(\tau)\mathbf{q}\ +\ \mathbf{B}(\tau)\mathbf{q}\ +\ \mathbf{C}(\tau)\right\}\ . \end{aligned}$$

Replacing (3.26) in the term to be maximised leads, after some elementary algebra, to the quadratic polynomial:

$$\begin{aligned} P(e) = \frac{1}{2}\ e^2\ \mathbf{b}_1^t\mathbf{A}(\tau)\mathbf{b}_1\ &+\ e\left(\mathbf{b}_1^t\mathbf{A}(\tau)(\mathbf{F}(\tau)\mathbf{x}+\mathbf{g}(\tau)) + \mathbf{B}(\tau)\mathbf{b}_1\right) \\ &+\ \frac{1}{2}(\mathbf{F}(\tau)\mathbf{x}+\mathbf{g}(\tau))^t\mathbf{A}(\tau)(\mathbf{F}(\tau)\mathbf{x}+\mathbf{g}(\tau)) \\ &+\ \mathbf{B}(\tau)(\mathbf{F}(\tau)\mathbf{x}+\mathbf{g}(\tau)) + \mathbf{C}(\tau)\ . \end{aligned} \tag{3.27}$$

The leading term is $\xi(\tau)$ (Definition (3.17)). If $\xi(\tau) > 0$, there is no finite e maximising this expression, so that the optimal control does not exist and the value function is infinite. In practical terms, player 1 believes she can increase her payoff beyond any finite value.

If $\xi(\tau) < 0$, the optimal value of e is e^* such that $P'(e^*) = 0$. Therefore,

$$e^* = -\frac{1}{\xi(\tau)} \left(\mathbf{b}_1^t \mathbf{A}(\tau)(\mathbf{F}(\tau)\mathbf{x} + \mathbf{g}(\tau)) + \mathbf{B}(\tau)\mathbf{b}_1 \right) . \tag{3.28}$$

This corresponds to an optimal state vector at the first stage given by $\mathbf{q} = \Phi(\tau)\mathbf{x} + \Psi(\tau)$, with the matrices of Definitions (3.18) and (3.19).

Plugging (3.28) into (3.25) gives the expected result, after some elementary algebra. □

Notice that the case where $\xi(\tau) = 0$ for some τ is inconclusive without further analysis. Either the first order term in e in (3.27) is zero, in which case all controls are maximising, or it is not, and then no maximum exist (as in the case $\xi(\tau) > 0$). Both circumstances must be avoided in practice.

3.2.5.3 *Stationary and proportional conjectures*

We assume now that conjectures of players do not vary over time, that is: $\mathbf{F}^i(\tau) = \mathbf{F}^i$ and $\mathbf{g}^i(\tau) = \mathbf{g}^i$ for all τ (stationary conjectures). We assume also that the conjectures player i forms about other players are "proportional" to one another, in the sense that two families of real numbers u_j^i and v_k^i exist, such that:

$$f_{j,k}^i = u_j^i \, v_k^i . \tag{3.29}$$

This amounts to saying that player i thinks that all other players use an affine rule for constructing their control, and that all rules put the same *relative* weights v_k^i on the individual state of player k. The variation in the absolute values is contained in the multiplicative constants u_j^i, which depends on the player j. This technical assumption appears to be necessary in order to make any progress in the resolution of the n-player game. While it seems quite restrictive, it encompasses two practically important cases: two-player games, and all symmetric n-player games.

From the algebraic standpoint, this assumption implies that matrix $\mathbf{F}^i$ is of rank one, that is, of the form $\mathbf{F}^i = \mathbf{u}^i(\mathbf{v}^i)^t$ for some column vectors $\mathbf{u}^i$ and $\mathbf{v}^i$. The coordinate u_i^i is zero since players do not form conjectures about themselves. The following Lemma 3.3 states that under these conditions, players will actually mimic this conjectured behaviour and choose an affine feedback proportional to the conjectures. The class of affine reactions defined by some vector $\mathbf{v}^i$ has therefore some sort of internal consistency.

Another consequence of this particular form of conjectures is that player i will see her opponents as a group, and the analysis will be actually similar to that of a two-player game, for each player. This will become formally clear in Theorem 3.4.

Lemma 3.3 *Assume* $\mathbf{F}$ *to be of the form* $\mathbf{F} = \mathbf{u}\mathbf{v}^t$. *Then the optimal reaction of player 1 at state* τ *is of the affine state-feedback form:*

$$e(t) \;=\; \mu(\tau)\ \mathbf{v}^t\ \mathbf{x}(t)\ +\ c(\tau)\ ,$$

for some scalar values $\mu(\tau)$ *and* $c(\tau)$.

Proof. According to Lemma 3.2 and Definition (3.19) the optimal reaction of player 1 writes as:

$$e(\tau) \;=\; \mathbf{b}_1^t\left(\mathbf{I} - \frac{1}{\xi(\tau)}\mathbf{b}_1\mathbf{b}_1^t\mathbf{A}(\tau)\right)\ \mathbf{F}(\tau)\ \mathbf{x}(t)\ +\ c(\tau)\ ,$$

with $c(\tau) = \mathbf{b}_1^t\Psi$. Since $\mathbf{F}(\tau) = \mathbf{u}\mathbf{v}^t$, this can be rewritten as:

$$e(\tau) = \left[\mathbf{b}_1^t\left(\mathbf{I} - \frac{1}{\xi(\tau)}\mathbf{b}_1\mathbf{b}_1^t\mathbf{A}(\tau)\right)\ \mathbf{u}\right]\ \mathbf{v}^t\ \mathbf{x}(t)\ +\ c(\tau)\ .$$

The term between brackets is a scalar $\mu(\tau)$. □

Introduce the following scalar values, summarising the interactions of the two "players" (player 1 and a virtual player 2 representing the other players):

$$\ell_{11} \;=\; \mathbf{b}_1^t\mathbf{K}\mathbf{b}_1 \qquad \ell_{12} \;=\; \mathbf{b}_1^t\mathbf{K}\mathbf{u} \qquad \ell_{22} \;=\; \mathbf{u}^t\mathbf{K}\mathbf{u}\ , \tag{3.30}$$

$$z_1 \;=\; \mathbf{b}_1^t\mathbf{v} \qquad z_2 \;=\; \mathbf{u}^t\mathbf{v}\ . \tag{3.31}$$

Theorem 3.4 *Assume* $\mathbf{F}$ *to be of the form* $\mathbf{F} = \mathbf{u}\mathbf{v}^t$. *Define:*

$$A = \ell_{11}\ell_{22} - \ell_{12}^2$$
$$B = \ell_{11}z_2^2 + \ell_{22}z_1^2 - 2\ell_{12}z_1z_2$$
$$\delta = z_2\ell_{11} - z_1\ell_{12}\ .$$

Then for all $\tau \geq 0$,

i) $\xi(\tau) = \mathbf{A}(\tau)_{11}$ *is given by the following recurrence*

$$\xi(\tau+1) \;=\; \theta B + \ell_{11} \;-\; \theta\,\frac{\delta^2}{\xi(\tau)}\,, \tag{3.32}$$

starting from $\xi(0) = \ell_{11}$.

ii) The matrices $\mathbf{K}(\tau)$ *have the form:*

$$\mathbf{K}(\tau) \;=\; k(\tau)\,\mathbf{v}\,\mathbf{v}^t\,,$$

with $k(\tau)$ *a scalar such that* $\xi(\tau) = \theta z_1^2 k(\tau) + \ell_{11}$, *and given by the recurrence:*

$$k(\tau+1) \;=\; \frac{A + \theta B k(\tau)}{\ell_{11} + \theta z_1^2 k(\tau)}\,, \tag{3.33}$$

starting from $k(0) = 0$.

Proof. Statement *ii*) is proved by recurrence. When $\tau = 0$, we have $\mathbf{K}(0) = 0$, so that the properties holds since $k(0) = 0$.

Assume that the property is true for some τ. Then $\mathbf{K}(\tau)_{11} = k(\tau)(v_1)^2$. Therefore, the relation $\xi(\tau) = \ell_{11} + \theta z_1^2 \mathbf{K}(\tau)$ is a consequence of Definition (3.16), combined with the notation (3.30).

Computing the value of $\mathbf{A}(\tau)$ in Definition (3.16) from the recurrence value $\mathbf{K}(\tau)$, then replacing $\mathbf{A}(\tau)$ in (3.20), and rearranging gives (3.33) and proves *ii*).

Finally, changing the variable $k(\tau)$ into $\xi(\tau)$ in Recurrence (3.33) leads to (3.32) after simplifications. This proves *i*). □

The existence and the convergence of controls is therefore directly linked to the behaviour of the homographic recurrences (3.32) or (3.33).

The rest of the analysis assumes that the payoff of players is strictly concave in their own variable, which implies $\ell_{11} < 0$. Actually, since $\xi(0) = \ell_{11}$, this condition is necessary for the existence of a solution to the optimisation problem (Lemma 3.2).

Theorem 3.5 *The stationary feedback exists if and only if the following conditions are met:*

$$\theta B + \ell_{11} \;<\; 0 \tag{3.34}$$

$$(\theta B - \ell_{11})^2 \;+\; 4\theta A z_1^2 \;\geq\; 0\,, \tag{3.35}$$

$$A \geq 0 \quad \text{or} \quad \theta B - \ell_{11} \geq 0\,, \tag{3.36}$$

where A and B have been defined in Theorem 3.4.

Proof. Considering Recurrence (3.32), set:

$$f(x) \;=\; \theta B + \ell_{11} \;-\; \theta\,\frac{\delta^2}{x}\,.$$

If the sequence $\{\xi(\tau)\}$ converges, the limit $\xi(\infty)$ is such that $f(\xi(\infty)) = \xi(\infty)$. The solutions of $f(x) = x$ are the roots of the equation:

$$P(x) \;:=\; x^2 \;-\; (\theta B + \ell_{11})x \;+\; \delta^2\theta \;=\; 0\,.$$

The discriminant of this polynomial with respect to the variable x is:

$$(\theta B + \ell_{11})^2 \;-\; 4\theta\delta^2 \;=\; (\theta B - \ell_{11})^2 \;+\; 4\theta A z_1^2\,,$$

using the identity $\delta^2 = B\ell_{11} - Az_1^2$. This expression is the left-hand side of Condition (3.35). The existence of a limit $\xi(\infty)$ is therefore possible only if (3.35) holds. According to Lemma 3.2, the limit should be negative. The product of the two roots of $P(x)$ is $\delta^2\theta$, and is therefore positive. For the limit $\xi(\infty)$ to be negative, it is necessary that the sum of the two roots be negative, which is equivalent to (3.34). Conversely, if both these conditions hold, it is easy to see that $\{\xi(\tau)\}$ converges while staying negative if and only if $\xi(0) = \ell_{11} \le \xi_2$, where ξ_2 is the largest negative fixed point of f. It turns out that: $P(\ell_{11}) = -z_1^2\theta A$. Therefore, either $A \ge 0$ and ℓ_{11} is located between the two roots of P, or $A < 0$ and $P'(\ell_{11}) \le 0$ is a necessary and sufficient condition for ℓ_{11} to be smaller than the smallest root of P. But $P'(\ell_{11}) = \ell_{11} - \theta B$. Therefore, we must have Condition (3.36). □

Observe first that since $\ell_{11} < 0$, then all three conditions (3.34)–(3.36) hold for θ small enough. There exists therefore some value $\theta_{\max}$ (possibly infinite) such that convergence to a stationary feedback happens if and only if $\theta < \theta_{\max}$. The mapping f being convex on the interval $(-\infty, 0)$, when convergence occurs, the value of $\xi(\infty)$ is given by the *smallest* fixed point $x = f(x)$, that is, the smallest root of the polynomial $P(x)$.

Next, consider the case of a "Nash conjecture": players expect other players to stick to some constant value. In that case, the vector $\mathbf{v}$ is zero. Consequently, $z_1 = z_2 = 0$, and $B = \delta = 0$. If follows that the conditions of Theorem 3.5 are satisfied. The stationary feedback always exists in this case. Actually, it is obvious that the optimal feedback must be constant: the optimal play at each step is the best response to the conjectured move of the opponents. This was known from Theorem 3.1.

Finally, observe that if the instantaneous payoff function Π is concave, (or, equivalently, if the matrix $\mathbf{K}$ is definite negative), then

$$B = (z_2\mathbf{b}_1 - z_1\mathbf{u})^t\mathbf{K}(z_2\mathbf{b}_1 - z_1\mathbf{u}) < 0 .$$

Also, $A > 0$ because it is the determinant of a Gram matrix (see Horn and Johnson (1985), p. 47). It follows that conditions (3.34)–(3.36) hold necessarily. In a nutshell, *when the payoff function is concave, the stationary feedback always exists.*

When the conditions of Theorem 3.5 are satisfied, the sequence of value functions W_T converges, so that the (anticipated) payoff of player 1 is well defined. It is however common to ask that in addition the evolution of the state of the system under the stationary control be well behaved. We have:

Theorem 3.6 *When the stationary feedback is implemented, the evolution of the state anticipated by the player converges as time goes to infinity, if and only if:*

$$|\delta| < |\xi(\infty)| .$$

Proof. The anticipated trajectory of the state variable $\mathbf{x}(t)$ under the stationary feedback is the solution of the linear dynamical system

$$\mathbf{x}(t+1) = \Phi(\infty)\,\mathbf{x}(t) + \Psi(\infty) ,$$

starting from some initial state $\mathbf{x}(0) = \mathbf{x}_0$. This linear system converges for all possible initial state if and only if the spectral radius of the matrix $\Phi(\infty)$ is strictly less than one. We have found that $\Phi(\infty)$ is a matrix of rank one, of the form $\mathbf{z} \times \mathbf{v}^t$. Its unique non-zero eigenvalue is therefore $\mathbf{v}^t.\mathbf{z}$. Performing the computations, we find that $\mathrm{sp}(\Phi(\infty)) = |\mathbf{v}^t.\mathbf{z}| = |\delta/\xi(\infty)|$. Hence the result. □

3.2.5.4 *Feedback-consistent conjectures*

We now turn to the existence of a conjecture consistent with feedbacks. We assume that all players are symmetric: they have symmetric payoff functions, the same discount parameter θ, and formulate the same conjectures about a third player. In the notation of Definition (3.29), we have: $u^i_j = 1$ for all $i \neq j$, $v^i_k = v_k$ and $g^i_k = g$ for all i and k. Consistency will hold if the feedback calculated by player 1 coincides with this expectation. From Lemma 3.2, we know that the stationary feedback is of the form:

$$x_1(t+1) = e_1(t) = \mu(\infty)\,\mathbf{v}^t\,\mathbf{x}(t) + c(\infty) ,$$

whereas other players expect the evolution of player 1 to be:

$$x_1(t+1) \;=\; \mathbf{v}^t\,\mathbf{x}(t) \;+\; g\,.$$

The affine conjecture represented by vectors **v** and **g** is therefore feedback-consistent in the sense of Definition 3.4 if

$$\mu(\infty) \;=\; 1\,, \qquad c(\infty) \;=\; g\,.$$

The constant $\mu(\infty)$ can be expressed using $\xi(\infty)$, and $c(\infty)$ is part of the vector $\Psi(\infty)$. We obtain the result:

Theorem 3.7 *The conjecture is feedback-consistent if and only if:*

i) either $\mathbf{v} = 0$ *and* g *is the Nash equilibrium of the static game;*
ii) or the following conditions hold:

$$\theta \;=\; \frac{1}{z_1+z_2}\,\frac{\ell_{11}+\ell_{12}}{z_2(\ell_{11}+\ell_{12})-z_1(\ell_{12}+\ell_{22})}\,, \tag{3.37}$$

with $\theta > 0$ *satisfying Conditions* (3.34) *to* (3.36), *provided that*

$$\frac{z_1\delta}{\ell_{11}+\ell_{12}}\,(\ell_{11}+\ell_{22}+2\ell_{12}) \;\geq\; 0\,, \tag{3.38}$$

and if

$$g \;=\; \Psi(\infty)_1\,, \tag{3.39}$$

with $\Psi(\infty)$ *the limit of the sequence* $\Psi(\tau)$ *when* $\tau \to \infty$.

Proof. The sketch of the proof is as follows. Imposing $\mu(\infty) = 1$ amounts to imposing $\xi(\infty) = \delta/(z_1+z_2)$. Using the fact that $\xi(\infty)$ is a root of the polynomial $P(x)$ defined in the proof of Theorem 3.5, we obtain Condition (3.37). Next, it is necessary that the value of $\xi(\infty)$ above be the *smallest* root of the polynomial $P(x)$, in order for convergence to occur to the prescribed value. Condition (3.38) is necessary and sufficient for this to happen. Finally, Condition (3.39) is simply the consistence of the constant part in the affine conjecture. □

3.2.5.5 *Cournot's duopoly*

We now apply the results of the previous section to the familiar Cournot duopoly with symmetric, linear costs. The instantaneous profit function of firm i is given by:

$$\Pi^i = x_i\,(\alpha - \beta(x_i + x_j)) + bx_i + c \;=\; \beta x_i\,(\Gamma - (x_i + x_j)) + c\,,$$

for some constants $\beta, \Gamma, c > 0$. The multiplicative constant β can be set to 1 by changing the unit of profits. This profit function is an instance of the general form (3.12) with the matrices (for firm 1):

$$\mathbf{K}^1 \;=\; \begin{pmatrix} -2 & -1 \\ -1 & 0 \end{pmatrix}, \qquad \mathbf{L}^1 \;=\; (\Gamma\;\;0)\,, \qquad \mathbf{M}^1 \;=\; c\,.$$

The conjecture each firm has about the other can be summarised in the two components of vector $\mathbf{v} = (a, b)^t$. Also, $\mathbf{u} = (0, 1)^t$ and $\mathbf{g} = (0, g)^t$.

The characteristic values appearing in the analysis are:

$$\begin{array}{c} \ell_{11} \;=\; -\,2, \quad \ell_{12} \;=\; -\,1, \quad \ell_{22} \;=\; 0, \quad z_1 \;=\; a, \quad z_2 \;=\; b, \\ A \;=\; -\,1, \quad B \;=\; 2b(a-b), \quad \delta \;=\; a - 2b\,. \end{array}$$

Accordingly, Recurrence (3.33) is in this case:

$$k(t+1) \;=\; \frac{1 + 2\theta b(b-a)k(t)}{2 - \theta a^2 k(t)}\,.$$

The results are summarised in the following statements. Their complete proof is straightforward, and is left out of this monograph.

Lemma 3.8 *The stationary feedback exists under the condition:*

$$\theta \;<\; \min\left\{\frac{1}{b^2}, \frac{1}{(a-b)^2}\right\}.$$

Theorem 3.9 *The repeated static Nash equilibrium is the unique feedback-consistent equilibrium in the quadratic Cournot game.*

Proof. According to Theorem 3.7, and letting aside the repeated static Nash solution, coincidence of conjectures and feedbacks may occur only if:

$$\theta \;=\; \frac{3}{3b-a}\,\frac{1}{a+b} \;\geq\; 0\,, \quad \text{and} \quad a\,(a-2b) \;\geq\; 0\,.$$

The conditions of convergence of the feedback of Theorem 3.5 write here as:

$$(a+3b)\,(a-2b) \;<\; 0 \quad \text{and} \quad a\,(a-5b) \;\geq\; 0\,,$$

for (3.34) and (3.36), Condition (3.35) being necessarily satisfied. It turns out that these four conditions are not compatible. There is therefore no other consistent equilibrium than the repeated Nash equilibrium of the static game. □

Similar computations allow to state that the same conclusion applies to the symmetric Bertrand duopoly.

3.2.5.6 *A distance game*

We conclude with the analysis of a distance game. This example is a typical case of a game with concave payoff functions. Player 1 wishes to minimise her distance to point $(1, 0)$ whereas player 2 wishes to minimise her distance to point $(0, 1)$. The instantaneous payoff function of player i is therefore given by:

$$\Pi^i = -(x_i - 1)^2 - x_j^2 = -x_i^2 - x_j^2 + 2x_i - 1 .$$

It is easy to see that the one-shot game has the following elements: the Nash equilibrium is $x_1 = x_2 = 1$, and the symmetric Pareto outcome is $x_1 = x_2 = 1/2$.

Let us now turn to the computation of the optimal feedback, given some affine conjecture. The matrices of the general form (3.12) are then:

$$\mathbf{K}^1 = \begin{pmatrix} -2 & 0 \\ 0 & -2 \end{pmatrix} , \qquad \mathbf{L}^1 = \begin{pmatrix} 2 & 0 \end{pmatrix} , \qquad \mathbf{M}^1 = -1 .$$

Again, the conjecture each player has about the other is described by the three vectors $\mathbf{v} = (a, b)^t$, $\mathbf{u} = (0, 1)^t$ and $\mathbf{g} = (0, g)^t$.

The characteristic values appearing in the analysis are now:

$$\begin{aligned} \ell_{11} = -2, \quad \ell_{12} = 0, \quad \ell_{22} = -2, \quad z_1 = a, \quad z_2 = b, \\ A = 4, \quad B = -2(a^2 + b^2), \quad \delta = -2b . \end{aligned}$$

Accordingly, Recurrence (3.33) is in this case:

$$k(t+1) = \frac{4 - 2\theta(a^2 + b^2)k(t)}{-2 + \theta a^2 k(t)} .$$

The conditions of convergence of this recurrence are always met, since the matrix $\mathbf{K}$ is definite negative.

According to Theorem 3.7, if a feedback-consistent equilibrium exists, Conditions (3.37) and (3.38) should hold, that is:

$$\theta = \frac{1}{b^2 - a^2} > 0 , \quad \text{and} \quad 4ab \leq 0 .$$

For any given discount factor θ, there exists an infinity of pairs (a, b) fulfilling these conditions:

$$\text{if } a \geq 0,\ b = -\sqrt{a^2 + \frac{1}{\theta}}, \quad \text{or if } a \leq 0,\ b = \sqrt{a^2 + \frac{1}{\theta}}\ . \tag{3.40}$$

For each of them, computations show that there exists a unique value of g satisfying Condition (3.39):

$$g \ = \ \frac{a^2 - b^2 + b}{b - a}\ . \tag{3.41}$$

We have therefore proved that:

Theorem 3.10 *There exists an infinity of feedback-consistent equilibria in the distance game: the repeated static Nash equilibrium, and a continuum of equilibria parametrised by b according to equations* (3.40) *and* (3.41).

Continuing the analysis, we look more precisely at the steady state of the game: what is the behaviour of the sequence $x(t+1)$ in a situation of feedback-consistent conjectures? Looking back at the state evolution equation (3.23), it is found that:

$$x(\infty) \ = \ (\mathbf{I} - \Phi(\infty))^{-1}\ \Psi(\infty) \ = \ \frac{g}{1-(a+b)} \begin{pmatrix} 1 \\ 1 \end{pmatrix}\ .$$

After simplifications, the coordinates of this vector write as (in the case $a > 0$):

$$x_1(\infty) = x_2(\infty) \ = \ \frac{1 + \sqrt{a^2\theta^2 + \theta}}{1 + a\theta + \sqrt{a^2\theta^2 + \theta}}\ .$$

When $\theta \to 0$, this values goes to 1, whatever the value of a. The vector $(1,1)^t$ is the Nash equilibrium of the static game. However, for any nonzero value of θ, the limit of this quantity when $a \to \infty$ is $1/2$. The vector $(1/2, 1/2)^t$ is the symmetric Pareto outcome of the static game.

3.3 The model of Başar, Turnovsky and d'Orey

Başar *et al.* (1986) propose a model featuring interdependent countries. The purpose of the policy makers is to find the optimal tradeoff between the rate of inflation and the level of unemployment, in a dynamic context.

The model is a discrete-time, infinite-horizon game with two players. The instantaneous payoffs are quadratic, and the dynamics are linear. The goal of the paper is to compute and compare several types of equilibria: Nash, Stackelberg, and a new solution based on Bresnahan's consistent conjectural variations equilibrium for static games.

This new definition uses the fact that the dynamic game can be reduced to a static game at each time step, using the dynamic programming principle. Bresnahan's consistent equilibrium with constant conjectural variations is computed for each static game, and the value function obtained is propagated backwards in time. The fact that the consistent equilibrium can be computed owes to the fact that both the instantaneous payoffs *and* the value functions are quadratic. Computations are therefore similar to the cases studied in Section 1.4.

As in Friedman (1977), the problem is first solved with a finite time horizon T. Then, the limit of controls and value functions is sought when $T \to \infty$. The convergence is observed numerically: no theoretical convergence conditions have been found for this type of solution. Among the conclusions of the paper, it is shown that playing the new "step-by-step consistent" solution increases the payoff of both players with respect to playing the feedback Nash solution.

3.4 Conclusion

In this chapter, we have put together several concepts of conjectural equilibria taken from different papers. This allows us to draw a number of perspectives.

First, finding in which class of dynamic games the different definitions of Section 3.2.1 coincide is an interesting research direction.

As we have observed above, an equilibrium according to Definition 3.3 (control-consistent) is an equilibrium according to Definition 3.1 (state-consistent). We feel however that when conjectures are of the simple form (3.3), it may be that actions of the players are not observable, so that players should be happy with the coincidence of state paths. State-consistency is then the natural notion. The stronger control-consistency is however appropriate when conjectures are of the "complete" form. We have further observed that for repeated games, Definitions 3.1 and 3.3 coincide.

The problem disappears when consistency is considered, since the requirements of feedback-consistency (Definition 3.4) imply the coincidence of conjectures and actual values for both controls and states. Indeed, if the feedback functions of different players coincide, their conjectured state paths and control paths will coincide, since the initial state $\mathbf{x}_0$ is common knowledge.

Another issue is that of the information available to optimising agents. In the models of Section 3.2, agents do not know the payoff functions of their opponents when they compute their optimal control, based on their own conjectures. Computing an equilibrium requires however the complete knowledge of the payoffs. On the other hand, *verifying* that a conjecture is consistent requires less information. Weak control-consistency (and the weak state-consistency that could have been defined in the same spirit) is verified by the observation of the equilibrium path. The stronger state-consistency, control-consistency and feedback-consistency can be checked by computations based on the knowledge of one's payoff function and the conjectures of the opponents.

For the approach of Başar *et al.* (1986), agents need complete information to compute their controls. The concept is therefore subject to the same epistemic rationality problems that apply to static conjectural variations equilibria, as we have discussed in Chapter 1.

Finally, it is possible to discuss the multiplicity of equilibria in the different models presented in this chapter. The results obtained by Fershtman and Kamien (1985) and Laitner (1980) and those we have obtained following Friedman (1968) show that all three approaches may lead to such a multiplicity.

This multiplicity may be inherited from the multiplicity of solutions of infinite-horizon dynamic or repeated game problems. The method we have followed for computing feedback-consistent equilibria performs a double selection. First, it selects the solutions of the infinite-horizon problem that are limits of solutions of the finite-horizon problem. Second, it imposes a stronger consistency requirement. Apparently, this does not eliminate the possibility of a continuum of equilibria.

So far, only the case of a duopoly has been solved with the three methods. In this specific case, there is a single feedback-consistent equilibrium, whereas multiple equilibria appear with the other equilibrium concepts. Also, the results of this chapter are limited to some linear-quadratic games with a particular family of affine conjectures. Other cases should be investigated before drawing general conclusions.

In the next chapter, we develop other approaches, centred around the idea of "learning", that may circumvent both problems discussed above: the (in)completeness of the information available to agents, and the choice of an equilibrium in the case of multiplicity.

Chapter 4

Dynamic Conjectures, Incomplete Information and Learning

4.1 Introduction

This chapter presents three dynamic conjectural variations models that are meant to depart from the fully-informed fully-rational agent paradigm. A general and unified approach to modelling boundedly rational behaviours has not surfaced so far, and the conjectural variations framework has been suggested as one possibility; another one is the evolutionary model approach. There are relationships between those two seemingly different views. We give an example of such a relationship in Section 4.5. It is worth noting, from an epistemic point of view, that the materials presented in this chapter are presumably immune to all the criticisms previously mentioned against CVE models.

In all those models, economic agents have a limited knowledge and a limited rationality. In particular, they do not know the payoff functions of the other agents. They do however observe the outcomes of past actions. They form conjectures about other players, and this is captured by a belief function, which they use for optimising their payoffs. The models differ from one another in the specific ways agents form and update their conjectures. This in turn gives rise to different optimising behaviours with distinctive properties.

In Itaya and Dasgupta (1995)'s model, the agents' conjectures vary over time. At each point of time agents select a best response, given the conjecture they make as for the other players' "reaction". Actual best responses may or may not correspond to their conjectured reaction functions. Then, they adjust their conjectures according to the discrepancy between actual and conjectural best responses. The resulting process is similar to the familiar "cobweb" of best-response reactions converging to the Nash equilibrium.

However, in the present model, conjectures converge possibly to that of the static CCVE of the game. The description of the model allows us to propose alternative conjecture-learning schemes which share this property. This analysis is presented in Section 4.2.

The model of Friedman and Mezzetti (2002) is a discounted repeated game, where agents form conjectures about the response of the opponent to a variation of their strategy between two consecutive time periods. Agents have an optimal stationary feedback strategy, given this conjecture. The implementation of this feedback rule results in a dynamical system on strategy profiles. Under appropriate conditions, the strategy profile played at time t converges to some limit. In a Bertrand oligopoly situation, this limit corresponds to the CVE of the associated static game, when the discount parameter approaches one. We show that this property is always verified: for every discount factor, the stationary strategies correspond to a CVE with a well-chosen constant conjecture. This model is the topic of Section 4.3.

In the third model, we propose an original learning process in which agents form conjectures on the response of their opponents to a variation of their strategy with respect to a known reference point. Players optimise step-by-step their conjectured payoff, then revise the slope of their conjecture, taking into account the strategies played by the opponents in the previous step of the game. This results in a dynamic process on conjectures. In the duopoly models of Cournot and Bertrand, we obtain that firms' strategies tend to a cartel solution in the steady state of the processes, provided some conditions are met on payoff functions, on learning parameters and on the reference point. We develop this model in Section 4.4.

Finally, Section 4.5 briefly discusses recent developments in the comparison of CCVE and equilibria defined in the theory of evolutionary games. We summarise and conclude the chapter in Section 4.6.

4.2 Conjecture adjustment process

The paper of Itaya and Dasgupta (1995) introduces the idea of "dynamic conjectures": agents form conjectures on variations of the opponent, as in the static games of Chapter 1, but they have the ability to *revise* their beliefs as a function of the discrepancy between the actual conjectural variation deduced from the *observed* actions of the opponents, and their current conjectural variations. The actions of player j are, of course, interpreted by player i as *reactions* to her own variations. This adjustment principle has the potential to let emerge, in the long run, the *consistent* conjectural variations equilibrium of the corresponding static game, since, if convergence occurs, the steady-state is where conjectured reaction functions and conjectural best responses coincide.

We shall first describe the ideas and the results of Itaya and Dasgupta's analysis, discussing in the process several issues about the adjustment of conjectures and the use of observations. This discussion points at several alternatives for the construction of a dynamic model describing the adjustment of the conjectures. However, we leave largely open the analysis of these possibilities. We merely investigate the comparison of two adjustment processes in the case of quadratic payoff functions, by which we conclude the section.

4.2.1 *Itaya and Dasgupta's conjecture adjustment process*

Itaya and Dasgupta (1995) have devised a conjecture adaptation scheme within the context of voluntary contributions to a public good. The model is that of Section 1.4.3, page 26, but with heterogeneous agents. There are two individuals with preferences defined over the consumption of a private good and of a public good. The decision variable e_i stands for agent i's contribution to the public good. If her total budget is I_i, her consumption of the private good is $c_i = I_i - e_i$.

Individual i is endowed with a Cobb-Douglas utility function:

$$V^i(e_i, e_j) = (I_i - e_i)^{\alpha_i} \ (e_i + e_j)^{1-\alpha_i} \ ,$$

where $0 < \alpha_i < 1$, $i = 1, 2$. In the *static* conjectural equilibrium game, the conjectural best response function of player j is given by (see Section C.2, page 153):

$$\chi_j(e_i) \ = \ \frac{I_j(1-\alpha_j)(1+r_i) - \alpha_j e_i}{(1-\alpha_j)r_i + 1} \ .$$

Accordingly, Itaya and Dasgupta propose that the "right conjecture" for player i at time t should have been the value:

$$\hat{r}_j(t) \ = \ \frac{\mathrm{d}\chi_j}{\mathrm{d}e_i} \ = \ - \frac{\alpha_j}{(1-\alpha_j)r_i(t)+1} \ . \tag{4.1}$$

If $r_j(t)$ and $\hat{r}_j(t)$ differ, player i should revise her conjecture. Itaya and Dasgupta propose an update of the conjecture based on the discrepancy between the current value $r_j(t)$ and the ideal value $\hat{r}_j(t)$. The resulting dynamic adjustment process is:

$$\dot{r}_j(t) = \frac{\mathrm{d}r_j}{\mathrm{d}t}(t) = \mu_i \left(-\frac{\alpha_j}{(1-\alpha_j)r_i(t)+1} - r_j(t) \right) , \tag{4.2}$$

μ_i being a positive parameter controlling for the speed of adjustment.

Let us come back to the formula giving the ideal conjecture $\hat{r}_j$. Under this form, it appears that player i has substituted herself to player j and computed *her* conjectural best response function. Itaya and Dasgupta do not give details about the way this value can actually be obtained based on observations (incomplete information) rather than on a computation involving the knowledge of the payoff function of the opponent (complete information). However, as we proceed to show, the simple observation of the relative variations of the opponent's strategies does not seem to provide this information, which suggests that (4.2) cannot be built without the knowledge of the opponent's payoff function. Let us remember that player i thinks that her opponent reacts proportionally to her increments in strategies. Accordingly, it is reasonable to assume that an estimate of the proportionality factor should be given by the ratio of increments in the value of the strategies, observed at two close time instants:

$$\hat{r}_j(t+\delta t) \approx \frac{e_j(t+\delta t) - e_j(t)}{e_i(t+\delta t) - e_i(t)} .$$

In the limit $\delta t \to 0$, this amounts to defining:

$$\hat{r}_j(t) = \lim_{\delta t \to 0} \frac{e_j(t+\delta t) - e_j(t)}{e_i(t+\delta t) - e_i(t)} = \frac{\dot{e}_j(t)}{\dot{e}_i(t)} .$$

Coming back to the model of the voluntary contributions to public good, assume that the strategies played at time t are the values of the static CVE corresponding to the current conjectural variations $(r_1(t), r_2(t))$. Then, from the first order condition, the following relations hold:

$$e_j(t) = \frac{(1-\alpha_j)(1+r_i(t))I_j - \alpha_j \, e_i(t)}{(1-\alpha_j)r_i(t)+1} ,$$

for $i,j = 1,2$, $i \neq j$. We have seen that the value of $\hat{r}_j(t)$ should be $\dot{e}_j(t)/\dot{e}_i(t)$. Differentiating the preceding relation with respect to time, we obtain:

$$\dot{e}_i(t) = -\frac{\alpha_j \dot{e}_j(t)}{(1-\alpha_j)r_i(t)+1} - \frac{(1-\alpha_j)((1-\alpha_j)I_j - \alpha_j e_i(t))}{((1-\alpha_j)r_i(t)+1)^2} \dot{r}_i(t) .$$

Therefore, the ratio $\dot{e}_j(t)/\dot{e}_i(t)$ does not coincide with $\mathrm{d}\chi_j/\mathrm{d}e_i$ obtained in Equation (4.1).

Assume nevertheless that the value $\mathrm{d}\chi_j/\mathrm{d}e_i$ can be somehow observed, and that conjectural variations evolve according to Equations (4.2). Itaya and Dasgupta have established that:

Theorem 4.1 *If a CCVE with strictly positive quantities exists, then it is unique and locally stable for all speeds of adjustment.*

4.2.2 *Principles*

We now come back to the basic ideas appearing in the previous section, and try to generalise them.

Two agents are involved in repeated interactions in continuous time. Agent i forms a constant conjectural variation r_j about her opponent. Based on the observed behaviour of the opponent, agent i will revise this conjecture over time.

Let $\hat{r}_j(t)$ denote in general the conjecture that i thinks she should have used, based on her observations up to time t. She then updates her conjecture based on the discrepancy between the current value $r_j(t)$ and this ideal value $\hat{r}_j(t)$. This leads to a differential equation of the form:

$$\dot{r}_j(t) = \mu_i \left(\hat{r}_j(t) - r_j(t)\right) , \tag{4.3}$$

with $\mu_i > 0$ the speed of adjustment.

Now, the issue is to compute the value $\hat{r}_j(t)$, and to precisely identify what information is needed to do that. The first thing to do is to make precise what strategies are played (and observed) at each point in time t, given the conjectures $(r_1(t), r_2(t))$.

Since players optimise with variational conjectures, they must possess a reference (benchmark) strategy profile, from which the variations are considered. Two possibilities appear immediately. Either the benchmark strategy is fixed throughout the game, or it evolves over time. As the first possibility is the cornerstone of the (discrete-time) model of Section 4.4, it will not be further discussed here. When adapting the benchmark strategy, the simplest choice is to take the last observed strategy profile as benchmark for next decisions. Still in discrete time, this idea is exploited in the model proposed by Friedman and Mezzetti (2002), which is exposed in Section 4.3. However, in continuous time, the problem is more delicate.

One naive reasoning is the following. Having observed the current strategy profile $(e_i(t), e_j(t))$, and forming the conjecture $r_j(t)$ about her opponent, player i should play "next" as optimal value the conjectural best response:

$$e_i(t + \delta t) \;=\; \chi_i(e_j(t)) \;.$$

Here, the value of δt is arbitrarily small. So, if the function $e_i(t)$ is continuous in time, player i should actually pick $e_i(t)$ such that

$$e_i(t) \;=\; \chi_i(e_j(t)) \;. \tag{4.4}$$

If both players do this way, the conjectural equilibrium with conjectures $(r_1(t), r_2(t))$ will result. The fallacy in this reasoning is that at time t, player i cannot observe $e_j(t)$ before playing $e_i(t)$. This is again the "double Stackelberg" paradox.

A way to avoid this problem is to consider that strategies are adjusted over time, much in the same way as conjectures. Since player i thinks she should have played according to Equation (4.4), she may decide to adjust her strategy according to

$$\dot{e}_i(t) \;=\; \frac{\mathrm{d}e_i}{\mathrm{d}t}(t) \;=\; \nu_i\left(\chi_i(e_j(t)) \;-\; e_i(t)\right) \;, \tag{4.5}$$

where ν_i is some speed of adjustment. Observe that if this dynamic system converges,[1] it must be to the CVE corresponding to the limit conjectures $(r_1(\infty), r_2(\infty))$. But this time, there is no observation paradox involved. When the adjustment speed ν_i increases, convergence occurs presumably faster, provided it occurs at all. In the limit, the strategy played at each instant in time is indeed the conjectural equilibrium. Therefore, the "double Stackelberg" problem can be circumvented, provided the learning scheme (4.5) is stable enough.

We can now come back to the construction of an estimate of what may be the conjectural variation of one's opponent.

The first reasoning is that of the previous section. If player i observes increments of the opponent's strategy and computes the ratio with her own increments, she should obtain an estimate of (what she thinks is) the proportionality factor used by her opponent. As we have seen, this leads

[1] When conjectures are constant over time, conditions under which the dynamical system (4.5) is locally stable are readily established. They generalise those under which the Nash equilibrium is locally stable (see for instance Fudenberg and Tirole (1991), section 1.2.5).

to an estimate

$$\hat{r}_j(t + \delta t) \approx \frac{e_j(t + \delta t) - e_j(t)}{e_i(t + \delta t) - e_i(t)},$$

and, in the limit $\delta t \to 0$, to:

$$\hat{r}_j(t) = \lim_{\delta t \to 0} \frac{e_j(t + \delta t) - e_j(t)}{e_i(t + \delta t) - e_i(t)} = \frac{\dot{e}_j(t)}{\dot{e}_i(t)} . \tag{4.6}$$

It can be argued that the values $\dot{e}_i(t)$ can be deduced by the players from the observations. With this definition, Equations (4.3), (4.5) and (4.6) define a complete dynamical system with four variables $r_1(t), r_2(t)$, $e_1(t)$ and $e_2(t)$. We leave the analysis of the convergence properties of this system open for the time being.

An alternative possibility for deriving an estimate $\hat{r}_j$ is the following. Assume that the knowledge of both players includes the strategies played $e_i(t)$ as well as the fact that both play with conjectures, and assume that the values $r_k(t)$, $k = 1, 2$, are somehow observed. Then, if player i wishes her conjecture r_j to be consistent, she knows that Equation (1.14) (in Corollary 1.4, page 18) must hold:

$$(1 + r_i r_j)\ V^i_{ij}(e_i, e_j) + r_i\ V^i_{ii}(e_i, e_j) + r_j\ V^i_{jj}(e_i, e_j) = 0 . \tag{4.7}$$

Accordingly, she concludes from the observation of $e_i(t)$, $e_j(t)$ and $r_i(t)$ (the conjecture of player j) that her conjecture should have been:

$$\hat{r}_j(t) = - \frac{r_i(t)V^i_{ii}(e_i(t), e_j(t)) + V^i_{ij}(e_i(t), e_j(t))}{r_i(t)V^i_{ij}(e_i(t), e_j(t)) + V^i_{jj}(e_i(t), e_j(t))} . \tag{4.8}$$

This estimate involves only observed quantities, and the own payoff function of the player. When this payoff function is quadratic, the $V^i_{k\ell}$ are constant, and the observation of the strategy profile $(e_i(t), e_j(t))$ is not necessary. We study this case in more detail in Section 4.2.3.

Proceeding this way, we obtain again a complete dynamic system involving four variables, each of them representing some adjustment performed by a player *solely* based on observations and private information.

4.2.3 *Quadratic models*

Most of the two-player models in which constant CCVE can be computed, use quadratic payoff functions $V^i(e_i, e_j)$. In addition, they usually share the following feature: the second derivative V^i_{jj} is zero. Let us examine the properties of two of the three conjecture-learning procedure defined in the previous paragraph, in the quadratic setting.

When V^i is quadratic, the quantities V^i_{ii}, V^i_{jj} and V^i_{ij} do not depend on e_i and e_j. Therefore, Equation (4.7) can be solved directly for r_j, to provide the value of the variations that player i thinks she should conjecture on the part of player j, given the observation of player j's conjectural variations r_i. This yields:

$$\hat{r}_j \;=\; -\;\frac{r_i V^i_{ii} + V^i_{ij}}{r_i V^i_{ij} + V^i_{jj}}\;. \tag{4.9}$$

The adjustment process, in continuous time, gives then the differential system:

$$\dot{r}_j(t) = \mu_i\left(-\;\frac{r_i(t)V^i_{ii} + V^i_{ij}}{r_i(t)V^i_{ij} + V^i_{jj}} \;-\; r_j(t)\right)\;, \qquad j = 1,2.$$

The fixed points (r_1^*, r_2^*) of this dynamical system are the variations of the CCVE of the static game, solution of the system of equations:

$$r_j^* \;=\; -\;\frac{r_i^* V^i_{ii} + V^i_{ij}}{r_i^* V^i_{ij} + V^i_{jj}}\;, \qquad j = 1,2\;.$$

When does the differential system converge to this fixed point, at least locally? The local analysis around the fixed point (r_1^*, r_2^*) involves the gradient matrix:

$$\begin{pmatrix} -\mu_1 & \mu_1\,\dfrac{\partial \hat{r}_2}{\partial r_1} \\ \mu_2\,\dfrac{\partial \hat{r}_1}{\partial r_2} & -\mu_2 \end{pmatrix}\;,$$

evaluated at the point (r_1^*, r_2^*). The trace of this matrix is $-(\mu_1 + \mu_2) < 0$. According to the Routh-Hurwitz conditions (*e.g.* Horn and Johnson (1991)), it is necessary and sufficient for local stability that the determinant be positive. The determinant is here

$$\Delta \;=\; \mu_1\mu_2\left(1 \;-\; \frac{\partial \hat{r}_2}{\partial r_1}\frac{\partial \hat{r}_1}{\partial r_2}\right)\;, \qquad \text{with} \qquad \frac{\partial \hat{r}_j}{\partial r_i} \;=\; -\;\frac{V^i_{ii}V^i_{jj} - (V^i_{ij})^2}{(r_i^* V^i_{ij} + V^i_{jj})^2}\;.$$

In the usual case where $V^i_{jj} = 0$, we observe the simplifications:

$$\frac{\partial \hat{r}_j}{\partial r_i} \;=\; \frac{(V^i_{ij})^2}{(r_i^* V^i_{ij})^2} \;=\; \frac{1}{(r_i^*)^2}\;, \qquad \Delta \;=\; (1 - (r_1^* r_2^*)^{-2})\;.$$

The conclusion is then that the equilibrium point (r_1^*, r_2^*) is locally stable, for any value of the adjustment parameters, if and only if $(1-(r_1^* r_2^*)^{-2}) > 0$, that is, if and only if $|r_1^* r_2^*| > 1$.

Assume now that the conjecture adjustment process is the one proposed by Itaya and Dasgupta. We then have:

$$\hat{r}_j = \frac{d\chi_j}{de_i} = -\frac{r_i V_{ii}^j + V_{ij}^j}{r_i V_{ij}^j + V_{jj}^j} . \tag{4.10}$$

This value of $\hat{r}_j$ is also obtained by solving player j's consistency equation, that is:

$$(1 + r_i r_j)\, V_{ij}^j + r_j\, V_{jj}^j + r_i\, V_{ii}^j = 0 .$$

The adjustment process, is now driven by the differential system:

$$\dot{r}_j(t) = \mu_i \left(-\frac{r_i(t) V_{ii}^j + V_{ij}^j}{r_i(t) V_{ij}^j + V_{jj}^j} - r_j(t) \right) .$$

The fixed points of this system are also the variations of the CCVE. The local analysis around the point (r_1^*, r_2^*) involves the same gradient matrix, with the difference that now:

$$\frac{\partial \hat{r}_j}{\partial r_i} = -\frac{V_{ii}^j V_{jj}^j - (V_{ij}^j)^2}{(r_i^* V_{ij}^j + V_{jj}^j)^2} .$$

Assuming $V_{ii}^j = V_{jj}^i = 0$, we have:

$$r_j^* = -\frac{V_{ij}^j}{r_i V_{ij}^j + V_{jj}^j} , \qquad \text{and} \qquad \frac{\partial \hat{r}_i}{\partial r_j}(r_j^*) = -(r_j^*)^2 .$$

The conclusion is then that the equilibrium point (r_1^*, r_2^*) is locally stable, for any value of the adjustment parameters, if and only if $|r_1^* r_2^*| < 1$.

In conclusion, the adjustment processes resulting from the two points of view admit the same stable points, which are the pairs of conjectural variations of the CCVE of the game (which may or may not exist, depending on the parameters). However, both adjustment processes converge under opposite conditions. This last property is easily explained as follows. The consistency equation (4.7) can be written as $C^i(r_i, r_j) = 0$. In the first adjustment procedure, based on Equation (4.9), player i uses as estimate of the conjectural variations a function $\hat{r}_j(r_i)$ such that $C^i(r_i, \hat{r}_j(r_i)) = 0$. In the second one, based on Equation (4.10), this function is solution to

$C^j(\hat{r}_j(r_i), r_i) = 0$. Therefore, the functions $(\hat{r}_j, \hat{r}_i)$ used in the first process are *inverse* of the ones used in the second. The stability condition requires that the product of the slopes of these functions be, in absolute value, less than unity. The slope of the inverse of a function is the inverse of its slope. Consequently, the products of the two slopes, in the two adjustment processes, are inverse of one another, which explains the result.

4.3 The model of Friedman and Mezzetti

The paper of Friedman and Mezzetti (2002) features a repeated game: agent i, $i = 1, \ldots, n$ seeks to optimise her total payoff $\sum_{t=1}^{\infty} \theta^{t-1} \Pi^i(e(t))$, with $e(t)$ the profile of strategies played at time t, and $\theta \in (0,1)$ a discount parameter.

Agents are endowed with a particular belief about intertemporal reactions of the opponents. Specifically, agent i believes that her most recent change in strategies (that is, $e_i(t) - e_i(t-1)$) will induce a change in the strategy of the other agents. The strategy that agent i expects agent j to choose at $t+1$ is given by:

$$e_j(t+1) = e_j(t) + r_{ij}\,(e_i(t) - e_i(t-1)), \tag{4.11}$$

where r_{ij} is the (constant variational) conjecture. This mechanism implements the idea that conjectural variations in the classical static sense should be expressed in a dynamic model. Notice that the benchmark strategy (see Chapter 1) used by player i at time t is $(e_i(t-1), e_j(t))$.

Two questions arise for this type of models. First: in which context does an optimal policy exist? Second: what is the relationship between this optimal policy and the conjectural variations equilibria in the corresponding static model?

The first issue is addressed using optimisation theory. Indeed, incorporating the conjecture in player i's game-theoretic decision problem produces a simpler optimisation problem: she has to choose the optimal strategy profile $\{e_i(t); t \geq 0\}$, considering (4.11) as the evolution equation of some state variable. The optimisation yields a sequence of single-period policies of the form $e_i(t) = \phi_t^i(e(t-1), e_i(t-2))$. It is therefore optimal for player i to use ϕ_1^i at time $t = 1$. Since player i does not know the true actions of player j, she *does not* commit to the plan $\{\phi_t^i; t \geq 1\}$, but rather observes the outcome of the first step, and applies again ϕ_1^i to state $e(1)$. As a result, the action of player i is a *stationary* feedback policy

$$e_i(t) = \phi_1^i(e(t-1), e_i(t-2)) \ , \qquad i = 1, \dots, n \ . \tag{4.12}$$

To prove the existence of the optimal policy, the method of Friedman (1977) is to consider the finite-horizon problem with horizon T, for which the optimal plan is found in the form of a recurrence, by backward induction. The problem is then to prove that the limit exists as T goes to infinity. When it does, it is then necessary to assess the convergence of the dynamical system defined by (4.12).

Friedman and Mezzetti apply this method to Bertrand and Cournot oligopolies. They derive the optimal policy functions, and give sufficient conditions for prices or quantities to converge to a steady state. Moreover the article draws precise links between the steady state of the dynamical system and static conjectural variations equilibria: steady state equilibria can range from complete cooperation to fully noncooperative play. In addition, it is proved that when the discount parameter θ goes to 1, the limiting price profile approaches the CVE of the static game with the constant conjectures r_{ij}.

This result can be extended in the following property:

Theorem 4.2 *Let $e_i^s(r;\theta)$ be a fixed point of the dynamical system* (4.12), *for a fixed vector of conjectures r. Let $e_i^c(r)$ be the conjectural variations equilibrium with the constant vector of conjectures r, for the associated static game.*

If there exists $e_i^s(r;\theta)$, then there exists a $e_i^c(\theta r)$, and conversely. If both are unique, then they coincide:

$$e_i^s(r;\theta) \ = \ e_i^c(\theta r) \ .$$

Proof. The proof proceeds along the lines of Friedman and Mezzetti (2002). Observe first that the recurrence (4.11) on the conjectured strategies of agent j boils down to:

$$e_j(t+1) \ = \ e_j(0) + r_{ij} \ (e_i(t) - e_i(-1)) \ . \tag{4.13}$$

The value $e_i(-1)$ is the strategy of agent i at time $t = -1$. Since decisions are taken here using increments of strategies, it is necessary to specify the initial conditions as a couple $(e_i(-1), e_i(0))$.

Since agent i conjectures that agent j plays as in (4.13), then she faces an optimisation problem with variables $e_i(t)$, $t = 1, 2, \ldots$:

$$\max \sum_{t=1}^{\infty} \theta^{t-1} \, \Pi^i \big(e_i(t), \{e_j(0) + r_{ij} \, (e_i(t-1) - e_i(-1))\}_{j \neq i}\big) \, .$$

The initial value $e_j(0)$ is assumed to be observed by player i. The first order condition for variable $e_i(t)$ is:

$$\begin{aligned} 0 = \theta^{t-1} \, & \Pi^i_i \, (e_i(t), \{e_j(0) + r_{ij}(e_i(t-1) - e_i(-1))\}_{j \neq i}) \\ & + \theta^t \sum_{j \neq i} r_{ij} \, \Pi^i_j \, (e_i(t+1), \{e_j(0) + r_{ij}(e_i(t) - e_i(-1))\}_{j \neq i}) \, . \end{aligned}$$

Assume that there exists a stationary solution $e^s_i(r; \theta) = e_i(t)$, $\forall t = -1, 0, 1, \ldots$, it must satisfy the condition:

$$0 \; = \; \Pi^i_i \, (e^s_i, \{e^s_j\}_{j \neq i}) \; + \; \sum_{j \neq i} \theta \, r_{ij} \, \Pi^i_j \, (e^s_i, \{e^s_j\}_{j \neq i}) \, .$$

According to Equation (1.18) page 21, this condition is the first order condition for a conjectural equilibrium with constant conjectures equal to $\tilde{r}_{ij} = \theta r_{ij}$. The proof of the converse is identical if $e^c_i(\theta r)$ exists. □

The convergence results are shown to continue to hold in a more general model with adaptive beliefs. Friedman and Mezzetti assume that the belief of firm i about firm j adjusts to follow the average past value of the ratio of firm j's to firm i's price changes. This rule is coupled to a threshold rule preventing firms to adjust their prices when the observed variation is too small. Indeed, such an event could be due to some random noise or some externality of small impact, not captured in the model. This belief adaptation scheme does not modify the local stability properties of the model. However, it should be observed that the set of possible limiting values of the conjectures has not been analysed yet.

To conclude this section, we highlight some interesting features of this model. First, agents base their actions on a reasoning that does not involve the knowledge of the payoff functions of their opponents. In such an incomplete information setting, agents are not even Bayesian, *i.e.* they do not know what the different states of the world could be, and they do not have probabilities to put on them. The logic is more that of procedural rationality: the construction of player i's decision problem is actually part of her decision problem. This construction may involve conjectures. Second, conjectures in this model are exogenous, and no mechanism to choose rationally this conjecture has been described yet. The belief adaptation procedure mentioned above does not appear to result in analytically computable limiting conjectures. Third, strategies converge to conjectural variations equilibria of the associated static game, provided the latter exists, and under some conditions on the discount rate and the magnitude of the conjectures. Finally, in the case of Bertrand oligopoly, such a convergence occurs for conjectures leading to equilibria ranging from fully cooperative to fully noncooperative.

4.4 A learning model for conjectures

Adapting the work of Friedman and Mezzetti, Jean-Marie and Tidball (2002) propose a discrete-time learning model where players also have a linear conjecture about the behaviour of other players. Players optimise step-by-step their conjectured payoff (which corresponds to letting the discount factor θ go to zero in the model of Friedman and Mezzetti), and try to learn the proportionality factor in the (linear) conjecture of their opponents; this is similar to the adjustment process of Itaya and Dasgupta.

First, as in Friedman and Mezzetti (2002), it is assumed that agents have beliefs about other agents, which lead them to think that they will play in a certain subset of the space of strategies. Also, we consider the simplest sets: straight lines. Second, the agents maximise their immediate payoff, in the spirit of the Best Response Dynamics of the learning models in games Fudenberg and Levine (1998). A motivation for this static behaviour on part of our agents would be that the absence of knowledge on their opponents' abilities and motivations leads them to distrust the future. Third, since agents are nevertheless repeating the game, we shall endow them with the possibility of revising their beliefs according to observations. This will result in a "learning" process where agents, at each step of the game, update the idea they have about the behaviour of other agents (their "conjecture").

The result of these modelling assumptions is a dynamic system on conjectures, as in Itaya and Dasgupta's model, and in the spirit of Section 4.2.2.

We are interested primarily in the convergence of this system. Indeed, although the agents are boundedly rational, they should wonder whether their beliefs have any substance. The model assumes that errors in the conjectures are part of the game, and that agents use observations to "try to do better next time". Yet, this optimism would not last if the observed errors were too large in magnitude and too frequent. In contrast, if the values of conjectures stabilise (that is, converge) as time passes, agents should be convinced that they were somehow right in their beliefs.

The analysis will therefore focus on the properties of the limiting strategy profiles. In particular, we shall consider the efficiency of the solution, and see whether it is possible that agents, through this mechanism, reach a Pareto optimum.

Two examples are then analysed: Cournot's oligopoly and Bertrand's duopoly. In these examples we obtain that agents move to Pareto optimality in the steady state of the processes, provided certain conditions are met on profit functions and on learning parameters.

4.4.1 *Principle*

Player i, $i = 1, \ldots, n$, seeks to maximise her instantaneous payoff $V^i(\cdot)$. The assumption is that players are not Bayesian: they know neither what the sets of strategies of their opponents are, nor their payoff functions. For the purpose of her maximisation, player i makes, about the behaviour of player j, the conjecture that

$$e_j = e_j^b + r_{ij}(e_i - e_i^b) \qquad r_{ij} \in \mathbb{R} . \tag{4.14}$$

The vector $e^b = (e_1^b, \ldots, e_n^b)$ is called the *reference* (or benchmark, in the terminology of Chapter 1) profile of strategies. In other words, player i assumes that other players will observe her deviation from her reference point e_i^b, and deviate from their own reference point e_j^b by a quantity proportional to this deviation.

Among the different possibilities for capturing the move of the opponent as a function of one's own move, *affine* functions have about the simplest form one can think of. An affine rule can always be written as in (4.14), that is, as a variation with respect to the reference point. The strictly proportional rule is a special case of this form, with the reference strategies $e_1^b = e_2^b = 0$.

The reference point is considered as exogenously given, and of common knowledge. The only adjustable quantities will be the coefficients r_{ij} of the affine conjecture. We discuss this assumption in Section 4.4.4.

The goal of player i is to maximise her payoff function taking into account the conjectures that she has made about the other players. She therefore solves the maximisation problem:

$$\max_{e_i} V^i(e_i, (e_j^b + r_{ij}(e_i - e_i^b))_{j\neq i}) \ .$$

If the solution of this problem exists and is unique, this results in a function $e_i = \phi_i(e^b; r_i)$, where $r_i = (r_{ij})_{j\neq i} \in \mathbb{R}^{n-1}$ is the vector of the conjectures that i makes about *other* players.

As usual, ϕ_i may be obtained by solving the first order conditions (see (1.18) page 21):

$$V_i^i(e_i, (e_j^b + r_{ij}(e_i - e_i^b))_{j\neq i}) + \sum_{j\neq i} r_{ij} V_j^i(e_i, (e_j^b + r_{ij}(e_i - e_i^b))_{j\neq i}) = 0 \ , \tag{4.15}$$

if these happen to be sufficient conditions for optimality.

After the play, the moves of the players are observed by everyone. In particular, player i observes that player j has played e_j. She concludes that if her conjectural scheme has any reality, her conjecture should have been r'_{ij} such that

$$e_j = e_j^b + r'_{ij}(e_i - e_i^b), \quad \text{that is,} \quad r'_{ij} \ = \ \frac{e_j - e_j^b}{e_i - e_i^b} \ .$$

If the interaction between actors is to be repeated, player i should revise her conjecture. In order not to trust too much the observed values (this information may be noisy, player j may have cheated or made mistakes...), the update of the conjecture is done through a standard smoothing as in Section 4.2, adapted to discrete time:

$$r''_{ij} \ = \ (1 - \mu_i) r_{ij} \ + \ \mu_i \, r'_{ij} \ .$$

where the learning parameter μ_i belongs to the interval $(0, 1]$.

The global evolution of the conjectures is obtained by embedding the individual optimisation scheme in a dynamic process. Let $e_j(t)$, $j = 1, \ldots, n$, be the values played at stage t, and $r_{ij}(t)$ be the values of conjectures just before playing stage t.

Since player i revises her conjectures at each step, we have:

$$r_{ij}(t+1) = (1 - \mu_i) r_{ij}(t) \ + \ \mu_i \ \frac{e_j(t) - e_j^b}{e_i(t) - e_i^b}$$

with $e_i(t) = \phi_i(e^b; r_i(t))$. Observe the similarity of this expression with (4.3) of Itaya and Dasgupta's model.

What happens if all players apply the same rule simultaneously? In that case, the set of all conjectures will evolve according to the recurrence:

$$r_{ij}(t+1) = (1-\mu_i)r_{ij}(t) \ + \ \mu_i \, \frac{\phi_j(e^b; r_j(t)) - e_j^b}{\phi_i(e^b; r_i(t)) - e_i^b} \ . \tag{4.16}$$

Clearly, this is defined only when ϕ_i and ϕ_j have a value, and if $\phi_i(e^b; r_i(t)) \neq e_i^b$. In other cases, the new conjecture is undefined.

4.4.2 *General properties*

We report in this section some general properties of the recurrence (4.16) and its fixed points. They turn out to have a particular consistency property, and exhibit a close relationship with Pareto optima of the problem.

Consistency

Rewrite the evolution rule (4.16) in the more general form:

$$r_{ij}(t+1) \ = \ (1-\mu_i)r_{ij}(t) \ + \ \mu_i \, \frac{f_j(r_j(t))}{f_i(r_i(t))} \ , \tag{4.17}$$

for some continuous functions f_i, and let $\beta_i = f_i(r_i)$.

Assume that the recurrence (4.17) converges, that is: $r_{ij}(t) \to r_{ij}$ when $t \to \infty$. Since the functions $f_i(r)$ are continuous with respect to r, we have the fixed-point equation:

$$r_{ij} = (1-\mu_i)r_{ij} + \mu_i \frac{f_j(r_j)}{f_i(r_i)} \ .$$

Simplifying, and dividing by $\mu_i \neq 0$, we obtain the value of the fixed point:

$$r_{ij} = \frac{f_j(r_j)}{f_i(r_i)} \ = \ \frac{\beta_j}{\beta_i} \ . \tag{4.18}$$

The numbers r_{ij} have the following property:

$$r_{i_1 i_2} r_{i_2 i_3} \dots r_{i_p i_1} = 1 \qquad \forall i_1 \dots i_p .$$

In particular,

$$r_{ji} = r_{ij}^{-1} \qquad \forall i \neq j .$$

The vector $(r_{i1}, \dots, r_{i,i-1}, 1, r_{i,i+1}, \dots, r_{in})$ is the direction of a line of the space of strategy profiles, on which player i chooses her own strategy. This vector is proportional to $\beta = (\beta_1, \dots, \beta_n)$, according to (4.18). Therefore, in the limit, all players play on the same line passing through e^b with direction β. This result is a type of *consistency* between the beliefs of the players. If player i deviates by a value δ_i from the reference point, she will anticipate from player j a deviation $\beta_j \delta_i / \beta_i$. If player j *does* deviate from this quantity, she will anticipate the deviation δ_i from player i. This idea of what is a consistent state of affairs is different from Bresnahan's view of consistency, between conjectured reactions and best responses.

Since ϕ_i is continuous, $e_i = \phi_i(e^b; r_i)$ is the strategy played in the limit by player i.

Pareto optimality

A point y is Pareto-optimal if it solves the maximisation problem (see for instance Laffont (1988)):

$$\max_x \sum_i \lambda_i V^i(x) \,, \tag{4.19}$$

for some numbers $\lambda_i \geq 0$. A necessary condition for the Pareto-optimality of point y is then: there exist numbers $\lambda_i \geq 0$ such that the first order condition:

$$\sum_i \lambda_i \; V_j^i(y) \;=\; 0 \tag{4.20}$$

holds for all j. Introduce the gradient matrix $\nabla V(y) = (V_j^i(y))_{i,j}$. Then the vector $\lambda = (\lambda_1, \ldots, \lambda_n)$ is such that $\lambda \nabla V(y) = 0$. This implies that the matrix $\nabla V(y)$ is singular.

Consider now a point x, limit of Recurrence (4.16). If the first order derivatives of V^i are continuous, x satisfies Equation (4.15) for all i. Equivalently, multiplying by β_i and using (4.18): for all i,

$$0 \;=\; \beta_i V_i^i(x) \;+\; \sum_{j \neq i} \beta_j V_j^i(x) \;=\; \sum_j \beta_j V_j^i(x) \;.$$

This amounts to saying that the (column) vector β is such that $\nabla V(x) \beta = 0$. Again, this implies that $\nabla V(x)$ is singular.

We conclude that both the limits of the learning scheme and the Pareto-optimal strategy profiles x are such that $\nabla V(x)$ is singular. This makes them "likely" to coincide, but since the conditions we have used are not sufficient in general, verifications must be performed.

Concretely: if x is a limit point obtained by the convergence of the recurrence (4.16), then $\nabla V(x)$ is singular, and there exists a row vector γ

such that $\gamma \nabla V(x) = 0$. If this vector happens to be positive, then x is a candidate solution for the maximisation problem (4.19). In any case, the fact that x realises a maximum must be checked.

Conversely, this analysis gives us a geometric (necessary) condition for a Pareto strategy profile x to be a limit point of the recurrence: the line passing through x and the reference point e^b must be tangent to the iso-payoff sets of each player for the profit obtained at x.

Is is proved in Jean-Marie and Tidball (2002) that in Bertrand and Cournot oligopolies, convergence may indeed occur to Pareto outcomes, but that it may also occur with limits that satisfy Condition (4.20) without being Pareto-optimal.

The case of identical players.

In this section, we assume that there are $n \geq 2$ players which have the same payoff function, and therefore the same function ϕ_i:

$$\phi_i(e^b; r) \;=\; \phi(e^b; r)\ .$$

If in addition players share the same reference point $e_i^b = e^b$, then

$$r_{ij} \;=\; 1, \qquad \forall i \neq j \tag{4.21}$$

is a fixed point of the recurrence. Assume finally that the initial values $r_{ij}(0)$ are identical, and that $\mu_i = \mu$ for all i. In this situation, players are identical, in the sense that they have the same characteristics. However, they are unaware of that situation, and cannot use this information. Then we have the result:

Theorem 4.3 *In the case of identical players, the recurrence converges to 1, for any $0 < \mu < 1$ and any (common) initial condition.*

Proof. By recurrence, $r_{ij}(t)$ is the same value $r(t)$ for all i and j. Indeed, this is true for $t = 0$ by the symmetry assumptions, and if it holds for t, then

$$r_{ij}(t+1) \;=\; (1-\mu) r_{ij}(t) \;+\; \mu \times 1 \;=\; (1-\mu) r(t) \;+\; \mu\ .$$

The sequence $r(t+1) = (1-\mu) r(t) + \mu$ clearly converges to 1 for all values of $r(0)$ provided that $|1 - \mu| < 1$. □

4.4.3 *Results*

4.4.3.1 *Cournot's oligopoly*

We apply the conjecture learning procedure to a symmetric Cournot oligopoly game. We show first that there exist fixed points to the recurrence (4.16), and that the limiting strategy profiles correspond to Pareto solutions (Theorem 4.4). Then, we concentrate on the duopoly and discuss the stability of the limit point.

Assume that there are $n \geq 2$ firms. The strategies of firms are quantities.

The inverse demand function is assumed to be a linear function of the total quantity $E = e_1 + \ldots + e_n$ and the production cost is also linear in e_i. Therefore, profits are:

$$V^i(e_i, e_{-i}) = (\alpha - \beta \sum_j e_j)e_i - (be_i + c) = \beta e_i(\Gamma - \sum_j e_j) - c$$

where α, β, b and c are positive constants, and $\Gamma = (\alpha - b)/\beta$ is assumed positive as well.

The optimisation program of firm i, given the vector of conjectures $r_i = (r_{ij}; j \neq i)$, is:

$$\max_{e_i} \; \beta \, e_i \left(\Gamma - e_i - \sum_{j \neq i} (e_j^b + r_{ij}(e_i - e_i^b)) \right) - c \, .$$

After simplifications, the optimal strategy $\phi_i(e^b; r_i)$ is found to be:

$$\phi_i(e^b; r_i) \;=\; \frac{\Gamma - E^b}{2\sum_{k=1}^n r_{ik}} \;+\; \frac{e_i^b}{2} \, , \tag{4.22}$$

where we have used the natural convention $r_{ii} = 1$. This function is defined only when $\sum_{i=1}^n r_{ik} > 0$. Otherwise, second order conditions are not satisfied, and actually, the optimisation problem does not have a finite solution.

Existence and Pareto optimality of fixed points

We have the:

Theorem 4.4 *The unique fixed points of Recurrence* (4.16) *are* $r_{ij} = e_j^b / e_i^b$. *The corresponding strategies are Pareto optima.*

Proof. According to Section 4.4.2, possible solutions of the fixed point system (4.18) are of the form $r_{ij} = \beta_j / \beta_i$, with $\beta_i = \phi_i(e^b; r_i) - e_i^b$. Using

Equation (4.22), we obtain that

$$\beta_j = \beta_j \frac{\Gamma - E^b}{2\sum_i \beta_i} - \frac{e_j^b}{2},$$

and then by adding these equations, that $\sum_j \beta_j = (\Gamma - 2E^b)/2$, so that:

$$\beta_j = \frac{e_j^b}{2} \frac{\Gamma - 2E^b}{E^b}$$

Finally, the only fixed point of the recurrence on conjectures is: $r_{ij} = \beta_j/\beta_i = e_j^b/e_i^b$. Using these values in the function ϕ_i gives the steady state quantities

$$e_i = \frac{\Gamma}{2} \frac{e_i^b}{E^b}.$$

These verify: $\sum_{i=1}^n e_i = \Gamma/2$, and it is well known that for this symmetric oligopoly, the set of Pareto solutions e_i is precisely given by this equation. Therefore, the steady state quantities are always Pareto efficient. The learning procedure selects among the Pareto outcomes the only one with both quantities proportional to that of the reference point. □

4.4.3.2 *Bertrand's duopoly*

We now apply this conjecture learning procedure to a symmetric Bertrand duopoly game. The strategies of firms are prices. The profit function is given by:

$$V^i(e_i, e_j) = (e_i - A)(y_0 - \alpha e_i + \beta e_j) - B$$

with A, B, α, y_0 positive constants, $\beta \in \mathbb{R}$, and $\beta \neq 0$. The best strategy is in this case:

$$\phi_i(e^b; r) = \frac{1}{2}\left(A + \frac{y_0 - r\beta e_j^b + \beta e_i^b}{\alpha - r\beta}\right),$$

and it corresponds to a maximisation only if $\alpha - r\beta > 0$ (see Section 1.4.2).

The results of the analysis reported in detail in Jean-Marie and Tidball (2002) can be summarised as follows:

- The point $(r_1, r_2) = (1, 1)$ is always a fixed point of Recurrence (4.16), and the limiting strategy profiles are Pareto optimal if and only if $|\beta| < \alpha$.

- For some values of the parameters β and e^b, there exist two other fixed points of the form $(r^*, 1/r^*)$ and $(1/r^*, r^*)$. The strategy profiles corresponding to these "spurious" values are not Pareto optimal.
- There is no value of β for which one of these three possible fixed points is locally stable for *all* values of the reference point e^b. The limit therefore depends on the reference point.
- If the reference point is the (unique) Nash equilibrium, then the unique fixed point is $r^* = 1$, and the recurrence is locally stable around this value.

4.4.4 *Comments and limitations*

The present model has obvious limitations, which should be discussed.

The most questionable issue is the origin of the reference point, given *a priori*, and fixed during the game. This reference point is made necessary by the "variational" choice of the conjecture. One may think of it as fixed by some authority, as for instance the initial value of some pricing process, or as a regulation, from which undisciplined players try to depart. The reference point may also be the outcome of some internal analysis of the firms, for instance a Nash equilibrium. Also, firms may have announced objectives some time in advance, and comparisons are made with respect to this plan.

We have not considered the possibility for firms to revise their beliefs about the reference point, but this is clearly a research direction that is worth exploring. For instance, if the recurrence exhibits some instability, this should prompt players to realise that their reference point is not adequate. Also, the results obtained for this model show that the existence of an outcome of the learning process, and its value, generally depend on the reference point. The choice of a proper one is therefore important. An evolution of the present model will be to have the reference point move in time, according to some reasonable update rule. For instance, using the last observed value as reference point, like in Friedman and Mezzetti's model, is the obvious starting point of such a future study. Since the conjecture of players involves a reference point *and* the conjectural variation, models where the learning bears on both parameters at the same time should also be developed.

Concerning the optimisation problem, and from an economic standpoint, we should rather set a *constrained* maximisation problem, in order to ensure that conjectured values lie in a reasonable domain, and that the individual optimisation program always results in a well defined value.

Finally, there is the problem that the recurrence (4.16) is not always defined, either because the functions r_i and ϕ_j have a restricted domain of definition, or when $\phi_i(e^b; r_i(t)) = e_i^b$.

The issue of the domain of definition can be addressed by restricting the range of variables, as discussed above. Concerning the possibility that $\phi_i(e^b; a_i(t)) = e_i^b$, if at the same time $\phi_j(e^b; a_j(t)) \neq e_j^b$, player i has to conclude that her conjecture mechanism fails, and she must choose a value by other means. Among the ideas for correcting this problem, one may add a rule by which conjectures do not change if $\phi_i(e^b; a_i(t))$ is too close to e_i^b. This convention is used in Friedman and Mezzetti (2002). It has the advantage of making the reference point a fixed point of the process, which remains stable as long as no player is wandering too far from it. As discussed earlier, the disadvantage is that the learning stops with an unpredictable value. Another possibility is to change the reference point, see above.

4.5 Evolutionary games and consistent conjectures

In this section, we briefly mention recent models establishing a link between the theory of evolutionary games (see *e.g.* Weibull (1995)), and consistent conjectural variations equilibria for static games.

The papers of Dixon and Somma (2001) and Müller and Normann (2002) both consider Cournot's duopoly with quadratic costs, with a profit function of the form:

$$V^i(e_i, e_j) = (a - be_i - \tau e_j)\, e_i - \frac{c}{2}\, e_i^2 \,.$$

Both conclude by direct computations that if $c > 0$, the unique *Evolutionary Stable Strategy* (ESS) of the game is the consistent conjecture of the static game. When costs are linear ($c = 0$), Dixon and Somma (2001) conclude that no ESS exists, although the consistent conjectural equilibrium does exist: it corresponds in this case to the Bertrand solution (see Section 1.4.1).

Further research in this direction should establish whether this coincidence of concepts is limited to duopoly models.

4.6 Conclusion

The models reviewed in this chapter provide some attempts at describing boundedly rational agents. This is done by embedding the notion of conjectures in dynamic settings that account for the possibility of some forms of learning. Those approaches are also interesting in that they fulfil the need, expressed by many authors, to analyse conjectural variations in true dynamic models.

The discussion of the continuous-time conjecture adjustment process of Itaya and Dasgupta (1995) has led us to conclude that many possibilities appear, according to the information effectively available to agents. In particular, we have proposed a new learning scheme based uniquely on the observation of the strategies played by the opponents. However, we have left for future development the analysis of the properties of this scheme, and its application to standard economic models.

An intriguing particularity of the continuous-time models discussed so far is the absence of a *reference point.* Reference points are natural in conjectural models due to the *variational* nature of these conjectures: players have in mind a "benchmark" value for their opponents, observe the difference with actual values, and react accordingly. In contrast, the discrete-time models of learning studied in this chapter have this reference point as a common feature.

The reference point in Friedman and Mezzetti (2002) is set to the last observed value, whereas it remains fixed in the model developed in Section 4.4. Observe however that since the conjectural variation does not vary over time in Friedman and Mezzetti (2002), the reference point is actually constant (compare Equation (4.13) with (4.14)).

Another model with a reference point is studied by Kalai and Stanford (1985). This is a repeated game with stationary affine conjectures of the form (4.14). The optimal control of players taking this conjecture into account is computed. This optimal control coincides with the Nash equilibrium of the game only when the reference point is the conjectural equilibrium of the corresponding static game. It is further proved that the only case where this Nash equilibrium is subgame-perfect is when the conjectures are *constant*, that is, when the reference point and the conjectured strategies are the Nash equilibrium of the static game.

In Section 4.4 we have proposed a simple dynamic model of our own, with affine conjectures. This model singles out Pareto optima among the possible limit strategies of the process. In Cournot and Bertrand duopolies, outcomes can indeed converge to Pareto optima. We do have here an example in which a cooperative behaviour emerges out of individual optimisation procedures, when players' interactions are modelled by simple linear conjectures.

However, it is also shown that the convergence to efficient outcomes strongly depends on the value of the reference strategy: convergence may occur with a limit strategy which is not a Pareto optimum, or may not occur at all. In addition, these original results are mostly restricted to *local* convergence. Still, among this variety of behaviours, we have observed for both Cournot and Bertrand duopolies, that taking as reference point the Nash equilibrium yields a local convergence towards a Pareto solution.

A future direction of research should be the application of the learning scheme to other two-player games in order to confirm our findings about the emergence of Pareto strategies in the long run. Another direction will be to re-design the conjecture scheme in order to improve the stability of the convergence, and to avoid the drawbacks attached to the fixed reference point. It is important to realise that the Pareto efficiency of the limits is due to the simple geometric facts that conjectures of both players are aligned with one another, and tangent to their iso-payoff curves. This property should persist in other learning models as long as they keep the form (4.17). Finally, it will be interesting to endow the process with a more complex cost structure, and study it as a real dynamic game, as in the model of Friedman and Mezzetti (2002). Their results allow to expect stronger stability properties for such a model.

Chapter 5

Conclusion

By now the reader should have a rough idea of where the theory of conjectural variations equilibria currently stands. Where it may develop in a near future is an interesting question, about which we risk here a prospective answer.

The part of the literature that, as in Chapter 2, provides efforts to explicit the links between dynamic models and static shortcuts has probably achieved its maturity. We have the feeling that the exploration of the relationship of conjectures with trigger strategies, which are known as a possibility to increase cooperation on a non cooperative basis, has done its best in the theories of oligopoly and voluntary contributions to a public good. It is likely to be useful not only for theoreticians, but also econometricians who are working on those two archetypical situations, to help them to interpret empirical departures from Nash behaviours. The extension in other contexts would probably be a trivial task, with little chance of surprises. The relationship with feedback behaviours is more promising from this point of view. Moving away from the linear-quadratic framework, investigations should try to assess whether steady states of such games have indeed a general interpretation as conjectural variations equilibria, and if so, if the sign of conjectures can be derived using simple elements taken from the description of the game. This would give its full sense to the ideas we develop in Appendix B: assessing qualitatively the impact of conjectural variations (hence, if this research program is successful, of intertemporal forces) on outcomes and payoffs *without* having to compute the equilibria.

Concerning the models of Chapter 3, the prospects are unclear. Devising models incorporating the idea of consistency in dynamic games is still an important issue. But two major problems appear in the analysis of models proposed so far. First, the objection on the rationality of the behaviour of agents, put forth against static conjectural variations, applies also to some of the dynamic models of Chapter 3. Second, the analysis of some linear-quadratic repeated games has lead so far to conclude that there is a multiplicity of conjectural equilibria, or a total lack thereof, except for the strategy consisting in playing the static Nash equilibrium repeatedly.

On the other hand, the part of the theory that, as in Chapter 4, explores models with boundedly rational agents, is the most likely to develop in a near future. This research avenue stands as a possible approach for studying procedural rationality. The results described in this chapter point at several directions, involving other branches of game theory.

A first point to study is the relationship between conjectural variations and evolutionary games, which we have only touched upon at the end of Chapter 4. It is possible that evolutionary games can provide a reinterpretation of static conjectural variations, much in the same way as in Chapter 2.

Another avenue for research that we have left largely opened in Chapter 4 is on the behaviour of dynamic models obtained when agents are assumed to formulate some conjecture (which is a model of their rival's behaviour), and learn the parameters of this model based on observations. For instance, the examples that have been studied so far have used affine conjectures as a model of the rival. What if agents use more complicated conjectures? Some results of Chapter 4 single out Pareto optimal outcomes as principal stable points of the learning processes. This is at odds with most results in the theory of learning in games, which usually suggest that Nash equilibria (of the static game) should emerge as the effect of repeated interactions. Is this emergence of some form of cooperation linked to the specificity of conjectures in an essential way? This question is actually central within the field of machine learning, where the principles of conjectural variations have also been tested. As far as economic models are concerned, empirical studies may be necessary to guide the modeller in the maze of possibilities

Because conjectural variations equilibria will continue to appear in scientific works, it is also clear that economists would welcome an exploration of the foundations of Game Theory in relation with the idea of conjectures. From the discussion given in Chapter 1, static conjectural models are highly suspicious in situations of complete information and common knowledge unless they represent shortcuts. Some models of Chapter 3 are suspicious as well in that respect. Chapters 2 and 4 are so far, and for different reasons, immune to this criticism. Will future epistemic developments about game theory destroy also those last contents? Writing on the foundations of game theory Binmore (1992), summarises his view as follows: "In brief, I believe the foundations of game theory to be a mess". We hope the present monograph will not add too much to the mess.

Appendix A

Properties of Conjectural Equilibria

This annex is devoted to the theoretical properties of Conjectural Variations Equilibria, in static settings. First we show results of geometric nature connecting CVE to Iso-payoffs curves. We then exhibit several families of payoff functions which admit a Consistent CVE. We also show conditions under which a Nash equilibrium is a CCVE.

The discussion in this appendix is restricted to two-player games.

We shall use the following definition:

Definition A.1 (ISO-PAYOFF CURVES AND FUNCTIONS)

An *iso-payoff curve* for player i is a subset of strategy profiles that yield the same payoff value for this player. Formally, it is given by:

$$\{(e_i, e_j) \in \mathbb{E} \mid V^i(e_i, e_j) = C_i\} ,$$

for some payoff level C_i. In short, an iso-payoff curve is made of all the values for the rival's strategies for which player i gets the payoff C_i when she plays e_i.

An *iso-payoff function* of player i is implicitly given by the solution of the equation $V^i(e_i, e_j) = C_i$ with respect to e_j, the payoff level C_i and the variable e_i being given. We shall denote each such solution by $e_j = \mathcal{I}_j(e_i; C_i)$.

A.1 Iso-payoffs curves and conjectured reaction functions

In this section, we establish geometric relationships between Iso-payoffs curves, conjectured reaction functions and conjectural best response functions. Cornes and Sandler (1984) make use of some of these properties in their "graphical device", in order, for instance, to derive qualitative properties of best response functions (increasingness, convexity, ...) or compare the values of equilibrium strategies obtained with different conjectures.

Consider a benchmark strategy $e^b \in \mathbb{E}$ and the conjectured reaction curve $e_j = \rho_j^c(e_i; e^b)$ (See Section 1.3.4, page 10).

Consider $\phi_i(e^b)$ the set of best responses, that is, the solutions of the maximisation problem (1.25):

$$V^i(\phi_i(e^b), \rho_j^c(\phi_i(e^b); e^b)) = \max_{e_i} V^i(e_i, \rho_j^c(e_i; e^b)) =: C_i(e^b) , \qquad \text{(A.1)}$$

where $C_i(e^b)$ is the payoff obtained at the maximum. In this paragraph, we shall assume that this maximisation problem has a unique solution. In other words, the set $\phi_i(e^b) \subset E_i$ is reduced to a single element, and with an usual abuse of notation, we shall denote also by $\phi_i(e^b)$ this element. We shall also use C_i for $C_i(e^b)$ in the following.

Theorem A.1 *Assume that the functions ρ_j^c and $\mathcal{I}_j$ are differentiable with respect to e_i in the neighbourhood of the strategy e_i^b. Then, the curves $\rho_j^c(e_i; e^b)$ and $\mathcal{I}_j(e_i; C_i)$ are tangent at the strategy profile $e = (\phi_i(e^b), \rho_j^c(\phi_i(e^b); e^b))$, $\forall e^b$, under the condition $V_j^i(e) \neq 0$.*

Proof. Clearly, the two curves pass through the same point at $e_i = \phi_i(e^b)$, by definition of C_i in (A.1). By Definition A.1, the iso-payoff function $\mathcal{I}_j(e_i; C_i)$ satisfies, for all e_i:

$$V^i(e_i, \mathcal{I}_j(e_i; C_i)) = C_i .$$

When differentiating, for all e_i:

$$V_i^i(e_i, \mathcal{I}_j(e_i; C_i)) + V_j^i(e_i, \mathcal{I}_j(e_i; C_i)) \frac{\partial \mathcal{I}_j}{\partial e_i}(e_i; C_i) = 0 .$$

In particular when $e_i = \phi_i(e^b)$,

$$V_i^i(\phi_i(e^b), \mathcal{I}_j(\phi_i(e^b); C_i)) + V_j^i(\phi_i(e^b), \mathcal{I}_j(\phi_i(e^b); C_i)) \frac{\partial \mathcal{I}_j}{\partial e_i}(\phi_i(e^b); C_i) = 0 . \qquad \text{(A.2)}$$

On the other hand, the first order condition for the maximisation problem (A.1) is:

$$V_i^i(\phi_i(e^b), \rho_j^c(\phi_i(e^b); e^b)) + V_j^i(\phi_i(e^b), \rho_j^c(\phi_i(e^b); e^b)) \frac{\partial \rho_j^c}{\partial e_i}(\phi_i(e^b); e^b) = 0 . \qquad \text{(A.3)}$$

Under the condition on V_j^i, we have therefore from (A.2), (A.3) and the fact that $\mathcal{I}_j(\phi_i(e^b); C_i) = \rho_j^c(\phi_i(e^b); e^b)$, that

$$\frac{\partial \mathcal{I}_j}{\partial e_i}(\phi_i(e^b); C_i) = \frac{\partial \rho_j^c}{\partial e_i}(\phi_i(e^b); e^b) .$$

The two curves are therefore tangent. □

Observe that this property holds for every benchmark strategy e^b, not only at an equilibrium e^c.

This geometric situation of tangency between iso-payoff curves and a curve representing some constraint is typical of optimisation problems and frequently encountered in economics. This property generalises the well-known fact that the iso-payoff curve of a player reaches a maximum at her strategy of a Nash equilibrium. Indeed, the conjectured reactions are constant in that case; at the point of tangency, the iso-payoff curve is therefore horizontal.

The geometric situation of Theorem A.1 has led Cornes and Sandler (1984) to observe that the conjectural best response curve $\{(e_i, \chi_j(e_i))\}$ can be seen as a kind of "expansion path" obtained when varying the value of the payoff associated with the point of tangency.

The geometry of conjectures is illustrated in Figure A.1 for the model of voluntary contribution to a public good, introduced in Section 1.4.3 and developed in detail in Section C.2.

The top-left figure represents the situation as seen by player 1. The family of convex curves are the iso-payoff curves. The family of parallel dashed lines represents a constant variational conjecture (that is, affine conjectural responses $\rho_2^c(\cdot)$ for player 1), in this case with a slope $r < 0$. The isolated thick line represents the points of tangency: this is the conjectural best response of player 1: the function $\chi_1(\cdot)$.

The situation of player 2 is on the top-right figure. The iso-payoff curves are facing right. The parallel dashed lines are the conjectured responses $\rho_1^c(\cdot)$, and the thick black line the best response of player 2.

The superposition of both figures illustrates what happens when conjectures are consistent. In the bottom-left figure, it is seen that the best response of player 1 coincides with one conjectured response of player 2, and conversely. The four functions ρ_i^c and χ_i, $i = 1, 2$, appear therefore as two distinct curves. For a better readability, only the conjectures and the tangent iso-payoff curves are depicted. The thick cross materialises the Nash equilibrium of the game.

Finally, the bottom-right figure is the situation for the Nash equilibrium: variational conjectures are zero (dashed lines). Conjectural best responses (thick black lines) and conjectured reactions do not coincide here.

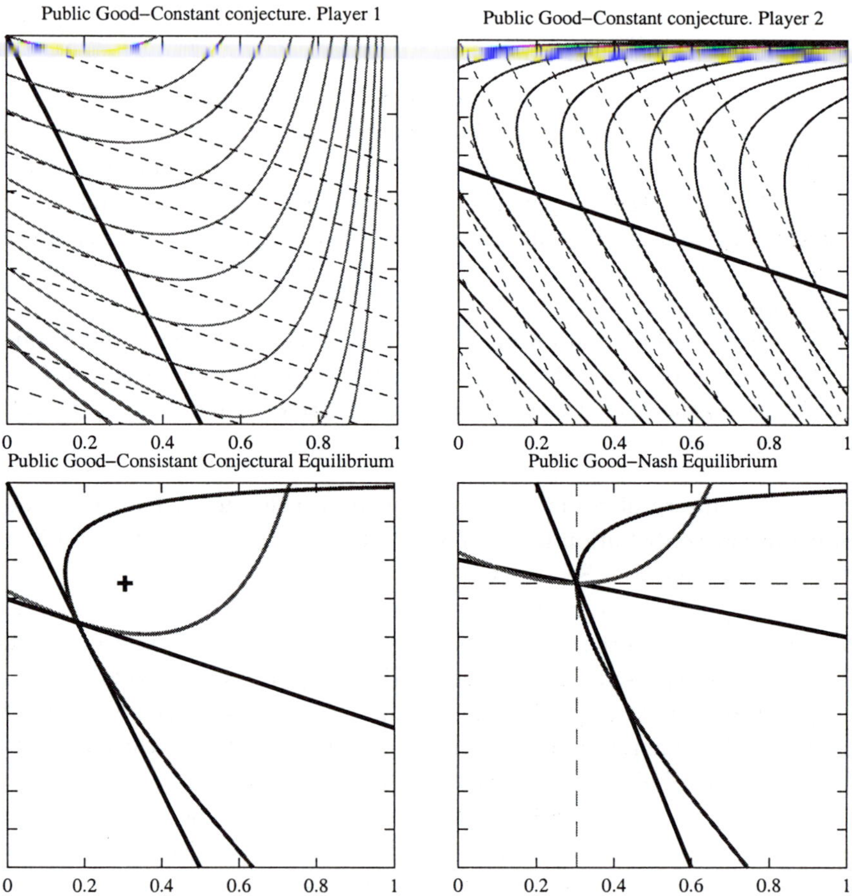

Fig. A.1 Conjectures, best responses and Iso-payoff curves in the model of voluntary contribution to a public good.

A.2 Families of payoff functions with consistent CVE

The next results aim at giving families of payoff functions V^i which admit consistent conjectures. Theorems A.2 and A.4 deal with general conjectures. Theorems A.5 and A.6 deal specifically with *constant* consistent conjectures.

We focus on conjectural variations of the form $r_i(e_j)$, leading to CCVE (Section 1.3.7). Contrary to the other types of equilibria reviewed in Chapter 1, it appears to offer the possibility of the existence of unique equilibria. Also, it covers the case of constant variational conjectures, which are the simplest conjectures one can think of, and practically the only ones used in the literature.

The first result establishes strong ties between conjectural best responses, iso-payoff curves and CCVE. Since these properties are stated for a CVE e^c, we shall use $\rho_i^c(e_j)$ for $\rho_i^c(e_j; e^c)$, with a slight abuse of notation.

Theorem A.2 *Let the strategy profile (e_1^c, e_2^c) and the variations $(r_1(e_2), r_2(e_1))$ form a consistent conjectural variations equilibrium. If the corresponding conjectured reaction functions ρ_1^c and ρ_2^c are inverse of one another (that is, $\rho_1^c \circ \rho_2^c = id$), then*

a/ there exist C_1 and C_2 such that for $i, j = 1, 2$:

$$\rho_i^c(e_j) \;=\; (\mathcal{I}_j)^{-1}(e_j; C_j), \qquad \forall e_j, |e_j - e_j^c| < \varepsilon ,$$

b/ any point of the curve $(e_1, \rho_2^c(e_1)) = (\rho_1^c(e_2), e_2)$ such that in addition $|e_i - e_i^c| < \varepsilon$, $i = 1, 2$ with the couple of conjectural variations $((\rho_1^c)', (\rho_2^c)')$, is a CCVE.

Conversely, if there exist C_1, C_2 such that Iso-payoff functions are inverse of one another, that is:

$$\mathcal{I}_1(\mathcal{I}_2(e_1; C_1); C_2) \;=\; e_1$$

for any e_1 in some open interval I_1, then for any $e_1 \in I_1$, $(e_1, \mathcal{I}_2(e_1; C_1))$ and $(\mathcal{I}_1{}'(e_2; C_2), \mathcal{I}_2{}'(e_1; C_1))$ form a CCVE.

Proof. Let (e_1^c, e_2^c) and $((\rho_1^c)', (\rho_2^c)')$ be a CCVE. Let us note I_i the interval $(e_i^c - \varepsilon, e_i^c + \varepsilon)$. Equation (1.12) writes: for e_2 in I_2,

$$V_1^1(\rho_1^c(e_2), e_2) + (\rho_2^c)'(\rho_1^c(e_2)) V_2^1(\rho_1^c(e_2), e_2) = 0 \; . \tag{A.4}$$

If $\rho_2^c \circ \rho_1^c(x) = x$ for all x, then

$$(\rho_1^c)'(e_2)\ (\rho_2^c)'(\rho_1^c(e_2)) \ = \ 1, \quad \forall e_2,$$

and multiplying (A.4) by $(\rho_1^c)'(e_2)$ gives, for $e_2 \in I_2$,

$$\begin{aligned} 0 &= (\rho_1^c)'(e_2)\ V_1^1(\rho_1^c(e_2), e_2) + V_2^1(\rho_1^c(e_2), e_2) \\ &= \frac{\mathrm{d}}{\mathrm{d}e_2} V^1(\rho_1^c(e_2), e_2)\ . \end{aligned}$$

Consequently, there exists a constant C such that

$$V^1(\rho_1^c(e_2), e_2) \ = \ C\ ,$$

and so $\rho_1^c(e_2) \ = \ (\mathcal{I}_2)^{-1}(e_2)$. A similar conclusion holds for player 1. This proves a/.

For b/, any point of the form $(e_1, \rho_1^c(e_2))$ satisfies Equation (1.12) and therefore it is a CCVE.

Conversely, if $\mathcal{I}_1(\mathcal{I}_2(e_1; C_1); C_2) = e_1$, that is:

$$(\mathcal{I}_1)^{-1}((\mathcal{I}_2)^{-1}(e_1; C_2); C_1) \ = \ e_1\ , \qquad \forall e_1\ ,$$

then taking $\rho_i^c = (\mathcal{I}_j)^{-1}$, every couple of points $(e_1, \mathcal{I}_2(e_1; C_1))$ and (ρ_1^c, ρ_2^c) satisfy Equation (A.4). □

Theorem A.2 exhibits the situation where the conjectured reaction functions of both players are consistent. Indeed, in this case player 1 believes that if she changes her strategy to e_1^m, player 2 will react with e_2^m such as to stay on some set $\mathcal{C}^1$ formed of strategy profiles $(e_1, \rho_2^c(e_1))$. But player 2 thinks that if she plays e_2^m, player 1 will react so as to stay in the set $\mathcal{C}^2$ formed of strategy profiles $(\rho_2^c(e_2), e_2)$. Since the sets are the same, these beliefs are consistent. According to the theorem, the equilibrium is consistent in the sense of Definition 1.5 if and only if the common conjectured reaction set is actually an Iso-payoff set for both players at the same time.

Two examples of this situation are known in the literature: the duopoly of Cournot with constant marginal costs (see Section C.1), and the public goods model (see Section C.2).

Observe that for a general CCVE, the conjectured reactions are *not* consistent with one another, but are consistent with conjectural *best responses.*

The next theorems aim at providing families of payoff functions V^i which admit (possibly non-constant) consistent CVE. As we have seen in Theorem 1.2 (page 15), a pair of consistent conjectural variations functions is the solution to a coupled difference-differential equation. There does not seem to exist a general method for solving this type of mathematical problem in closed form. Therefore, given payoff functions V^i, finding a consistent system of conjectures is a difficult problem. We circumvent this difficulty by considering the problem the other way round, and ask which payoff functions admit a given system of conjectures. Proceeding by analogy with known examples, we identify several such families of functions.

First, we have the classical change of payoff measure: if a CCVE exists for some payoff functions, it also exists for differentiable increasing transformations of them.

Theorem A.3 *Assume that* (e_1^c, e_2^c) *and the conjectural variations* (r_1, r_2) *form a CCVE for the couple of payoff functions* (V^1, V^2)*. Then they also form a CCVE for the couple* $(F^1(V^1), F^2(V^2))$*, for any differentiable functions* $F^i\colon \mathbb{R} \to \mathbb{R}$*, with* F^i *increasing.*

Proof. According to Theorem 1.3, the CCVE satisfies Equation (1.12). Since

$$\frac{\partial(F^i \circ V^i)}{\partial e_i}(e_i, e_j) \;=\; (F^i)'(V^i(e_i, e_j))\; V_i^i(e_i, e_j)\;,$$

then clearly (1.12) also holds when V^i is replaced by $F^i(V^i)$. The increasingness of F^i guarantees the preservation of maximality. □

Theorem A.4 *For* $i = 1, 2$*, let* $\varphi_i(u, v)$ *be a differentiable function from* $\mathbb{E}\times\mathbb{E}$ *into* $\mathbb{R}$*,* ρ_i^c *differentiable from* $\mathbb{E}$ *to* $\mathbb{E}$*,* g_i *and* h_i*, differentiable functions from* $\mathbb{E}$ *to* $\mathbb{R}$ *and* F^i*, differentiable from* $\mathbb{R}$ *to* $\mathbb{R}$*. If:*

$$V^i(e_i, e_j) \;=\; F^i\big(g_i(e_j)\; h_i(e_i)\; \varphi_i(e_i - \rho_i^c(e_j), e_j - \rho_j^c(e_i))\big), \qquad i = 1, 2 \tag{A.5}$$

then the points (e_i^c, e_j^c) *such that* $e_j^c = \rho_j^c(e_i^c)$ *and the conjectural variations* $(\rho_j^c)'(e_i), (\rho_i^c)'(e_j)$ *form a CCVE (if the appropriate maximality condition holds) in the following cases:*

i) $\varphi_i(0, v) = 0$ *for any* v*, and* $(F^i)'(0) = 0$*;*
ii) $\varphi_i(0, v) = 0$ *and* $\partial\varphi_i/\partial u(0, v) = 0$ *for any* v*;*
iii) $\varphi_i(0, 0) = 0$ *and* $\rho_i^c \circ \rho_j^c = id$*.*

Payoff functions that are sums of such forms enjoy the same property.

Proof. Let then V^i be defined above. When differentiating, we obtain:

$$\begin{aligned}
V_i^i(e_i,e_j) &= (F^i)'(x)g_i(e_j) \\
&\times\left[h_i'(e_i)\varphi_i(y_i,y_j)+h_i(e_i)\frac{\partial\varphi_i}{\partial u}(y_i,y_j)-h_i(e_i)(\rho_j^c)'(e_i)\frac{\partial\varphi_i}{\partial v}(y_i,y_j)\right] \\
V_j^i(e_i,e_j) &= (F^i)'(x)h_i(e_i) \\
&\times\left[g_i'(e_j)\varphi_i(y_i,y_j)-g_i(e_j)(\rho_i^c)'(e_j)\frac{\partial\varphi_i}{\partial u}(y_i,y_j)+g_i(e_j)\frac{\partial\varphi_i}{\partial v}(y_i,y_j)\right].
\end{aligned}$$

We have noted to simplify: $x = g_i(e_j)h_i(e_i)\varphi_i(e_i-\rho_i^c(e_j), e_j-\rho_j^c(e_i))$ and $y_i = e_i - \rho_i^c(e_j)$. Therefore:

$$\begin{aligned}
V_i^i(e_i,e_j)+(\rho_j^c)'(e_i)V_j^i(e_i,e_j) = (F^i)'(x)\Big[&(h_i'g_i+(\rho_j^c)'h_ig_i')\varphi_i \\
&+h_ig_i(1-(\rho_j^c)'(e_i)(\rho_i^c)'(e_j))\frac{\partial\varphi_i}{\partial u}\Big].
\end{aligned}$$

Finally, when evaluating the expressions at the point $e_i = \rho_i^c(e_j)$ using the notation $z_j = e_j - \rho_j^c\circ\rho_i^c(e_j)$, yields:

$$\begin{aligned}
&V_i^i(\rho_i^c(x_j),x_j)+(\rho_j^c)'(\rho_i^c(x_j))V_j^i(\rho_i^c(x_j),x_j) \\
&=(F^i)'(g_i(e_j)h_i(\rho_i^c(e_j))\varphi_i(0,z_j)) \\
&\quad\times\left[(h_i'g_i+(\rho_j^c)'h_ig_i')\varphi_i(0,z_j)+h_ig_i(1-(\rho_j^c)'\circ\rho_i^c(e_j)(\rho_i^c)'(e_j))\frac{\partial\varphi_i}{\partial u}(0,z_j)\right].
\end{aligned}$$

This last expression vanishes identically in all three cases. Indeed, if $\varphi_i(0,v)=0$, there remains:

$$(F^i)'(0)h_ig_i(1-(\rho_j^c)'\circ\rho_i^c(e_j)(\rho_i^c)'(e_j))\frac{\partial\varphi_i}{\partial u}(0,z_j)\ ;$$

hence cases *i*) and *ii*). On the other hand, if $\rho_i^c\circ\rho_j^c = id$, then on the one hand, $z_j = 0$, and on the other hand, $1-(\rho_j^c)'\circ\rho_i^c(e_j)(\rho_i^c)'(e_j)=0$. Hence the result. The function $\rho_i^c(e_j)$ is then solution of (1.10), or equivalently, (1.12) is true. We have therefore a CCVE. □

In case *iii*) of Theorem A.4, the conjectures are inverse of one another, so that we are in the situation of Theorem A.2. Indeed, for a strategy profile in the conjectured reaction set, say, $(e_i,\rho_j^c(e_i))$, the payoff of player i is:

$$V^i(e_i,\rho_j^c(e_i)) \;=\; F^i(g_i(\rho_j^c(e_i))h_i(e_i)\varphi_i(0,0)) \;=\; F^i(0)\,.$$

The payoff of both players is therefore constant along the conjectured reaction curve.

We now turn to *constant* conjectural variations. In that case, Corollary 1.4 (page 18) tells us that Equation (1.14):

$$(1 + r_i r_j)\ V^i_{ij}(e_i, e_j) + r_i\ V^i_{ii}(e_i, e_j) + r_j\ V^i_{jj}(e_i, e_j)\ =\ 0$$

holds for $e_i = \rho^c_i(e_j)$ and e_j close to the equilibrium. If one assumes that this equation holds actually for any pair of strategies (e_i, e_j) in the vicinity of the equilibrium, one gets a partial differential equation. All functions solutions of it satisfy the necessary condition (1.14), and are candidate for admitting a constant CCVE. This leads to the following result.

Theorem A.5 *Let r_i and r_j be two real numbers such that $r_i r_j \neq 1$. Let F^i and G^i be differentiable functions from $\mathbb{E}$ to $\mathbb{R}$, such that F^i admits a unique maximum f^i. Then if*

$$V^i(e_i, e_j) = F^i(e_i - r_i e_j)\ +\ G^i(e_j - r_j e_i)\ ,$$

the couple of strategies (e^c_1, e^c_2) with

$$e^c_i\ =\ \frac{f^i + r_i f^j}{1 - r_i r_j}\ ,$$

and the constant conjectures (r_1, r_2) form a CCVE.

Proof. First, we check that the optimisation problem of player i is well defined. Taking the strategy profile (e^b_1, e^b_2) as benchmark, and using her conjecture r_j, player i maximises:

$$V^i(e_i, e^b_j + r_j(e_i - e^b_i))\ =\ F^i(e_i(1 - r_i r_j) - r_i(e^b_j - r_j e^b_i))\ +\ G^i(e^b_j - r_j e^b_i)\ .$$

Since $1 - r_i r_j \neq 0$ and since f^i is the unique maximum of F^i, the equilibrium exists and is:

$$e_i\ =\ \frac{1}{1 - r_i r_j}\left(f^i\ +\ r_i(e^b_j - r_j e^b_i)\right)\ .$$

Next, we compute the conjectural best response of player i. From Theorem 1.1, it is obtained from the first order condition $V^i_i + r_j V^i_j = 0$, which is here equivalent to:

$$(F^i)'(e_i - r_i e_j)\ (1 - r_i r_j)\ =\ 0\ .$$

The conjectural best response function is therefore:

$$\chi_i(e_j)\ =\ r_i e_j\ +\ f^i\ .$$

The value of the CVE is obtained by solving the equations $e_i = \chi_i(e_j)$. This yields the values claimed for e_i^c. Obviously, $(\chi_i)' = r_i$, and the CVE is consistent. □

The case $r_i r_j = 1$ is not covered by this theorem. The reason is that in that case, the conjectures of both players are inverse from one another, and stronger results like Theorem A.2 apply. The general solution of the partial differential equation turns out to be $V^i = F^i(e_i - r_i e_j) + e_i G^i(e_i - r_i e_j)$. This solution belongs to the family of functional forms of Theorem A.4, *iii*).

Finally, we describe a parametric family of functions which possess, under technical conditions, constant CCVE. This family encompasses actually most examples known in the literature. We shall refer systematically to this general form when reviewing the results of the literature in Appendix C.

Theorem A.6 *Assume for each $i = 1, 2$, $V^i(e_i, e_j)$ is of the form*

$$V^i(e_i, e_j) = (a_i' e_i + a_i e_j + b_i)^{\mu_i} (c_i' e_j + c_i e_i + d_i)^{\nu_i} , \qquad \text{(A.6)}$$

where a_i, b_i, c_i, d_i, μ_i and ν_i are some constants. Then all constant pairs (r_1, r_2) such that r_1 is a real root of Equation (A.12) below, r_2 is given from r_1 by (A.11) and such that in addition Condition (A.9) below is satisfied, provide constant consistent conjectural variations, when the second order condition has checked.

Notice that according to Theorem A.3, the equilibria are the same for any function of V^i. For instance, we could take $\nu_i = 1$ without loss of generality (with $F^i(x) = x^{1/\nu_i}$). Or we could have, with $F(x) = \log(x)$,

$$V^i(e_i, e_j) = \mu_i \log(a_i' e_i + a_i e_j + b_i) + \nu_i \log(c_i' e_j + c_i e_i + d_i) .$$

Also, all functions of the form $A_i V^i + B_i$, $A_i > 0, B_i$ some constants, will have the same equilibria.

Using Theorem A.3 in conjunction with Theorem A.6 allows to find several examples of CCVE known in the literature. In particular, quadratic payoff functions are a special case of (A.6) obtained when $\mu_i = \nu_i = 1$.

Proof. Consider a constant conjecture r_j. Forming condition (1.10), one obtains:

$$\begin{aligned} 0 &= (a_i' e_i + a_i e_j + b_i)^{\mu_i - 1} (c_i' e_j + c_i e_i + d_i)^{\nu_i - 1} \qquad \text{(A.7)} \\ &\times \left(\mu_i (a_i' + a_i r_j)(c_i' e_j + c_i e_i + d_i) + \nu_i (c_i + c_i' r_j)(a_i' e_i + a_i e_j + b_i)\right) . \end{aligned}$$

A sufficient condition for (A.7) to hold is:[1]

$$\mu_i(a'_i + a_i r_j)(c'_i e_j + c_i e_i + d_i) \;+\; \nu_i(c_i + c'_i r_j)(a'_i e_i + a_i e_j + b_i) \;=\; 0\,, \tag{A.8}$$

for $i = 1, 2$. These two equations form a linear system with e_i and e_j as unknowns. It has a unique solution when the determinant is not zero. Some elementary algebra leads to the condition:

$$\begin{aligned}&(c_1\mu_1(a'_1 + a_1 r_2) + a'_1\nu_1(c_1 + c'_1 r_2))(c_2\mu_2(a'_2 + a_2 r_1) + a'_2\nu_2(c_2 + c'_2 r_1))\\ &\;\neq (c'_1\mu_1(a'_1 + a_1 r_2) + a_1\nu_1(c_1 + c'_1 r_2))(c'_2\mu_2(a'_2 + a_2 r_1) + a_2\nu_2(c_2 + c'_2 r_1))\,.\end{aligned} \tag{A.9}$$

At this stage of the analysis, Condition (A.9) gives a sufficient condition for the existence of a CVE.

When this condition is not met, the number of solutions (that is, of equilibrium points) is either 0 or an infinity.

From (A.7), it is deduced that conjectural best response have the form:

$$\begin{aligned}\chi_i(e_j) = &-\frac{c'_i\mu_i(a'_i + a_i r_j) + a_i\nu_i(c_i + c'_i r_j)}{c_i\mu_i(a'_i + a_i r_j) + a'_i\nu_i(c_i + c'_i r_j)}\, e_j\\ &-\frac{d_i\mu_i(a'_i + a_i r_j) + b_i\nu_i(c_i + c'_i r_j)}{c_i\mu_i(a'_i + a_i r_j) + a'_i\nu_i(c_i + c'_i r_j)}\,.\end{aligned} \tag{A.10}$$

The consistency conditions $r_i = \chi'_i(e_j)$, $i = 1, 2$, imply that (r_1, r_2) must be solution of the system:

$$\left\{\begin{aligned} r_1 &= -\frac{c'_1\mu_1(a'_1 + a_1 r_2) + a_1\nu_1(c_1 + c'_1 r_2)}{c_1\mu_1(a'_1 + a_1 r_2) + a'_1\nu_1(c_1 + c'_1 r_2)}\\ r_2 &= -\frac{c'_2\mu_2(a'_2 + a_2 r_1) + a_2\nu_2(c_2 + c'_2 r_1)}{c_2\mu_2(a'_2 + a_2 r_1) + a'_2\nu_2(c_2 + c'_2 r_1)}\,.\end{aligned}\right. \tag{A.11}$$

Replacing the second equation in the first one leads to a polynomial of the second degree, of which r_1 should be a real root:

$$\begin{aligned}&(r_1 c_1 a'_1(\mu_1 + \nu_1) + a'_1 c'_1\mu_1 + a_1 c_1\nu_1)(r_1(a_2 c_2\mu_2 + a'_2 c'_2\nu_2) + a'_2 c_2(\mu_2 + \nu_2))\\ &-(r_1 a_2 c'_2(\mu_2 + \nu_2) + a'_2 c'_2\mu_2 + a_2 c_2\nu_2)(r_1(a_1 c_1\mu_1 + a'_1 c'_1\nu_1) + a_1 c'_1(\mu_1 + \nu_1))\\ &= 0\,.\end{aligned} \tag{A.12}$$

---

[1] Conditions (A.7) and (A.8) are equivalent if none of the terms $a'_i e_i + a_i e_j + b_i$ or $c'_i e_j + c_i e_i + d_i$ are zero.

Conversely, since we have proceeded through sufficient conditions, any root of (A.12) provides a solution of (A.11). If Condition (A.9) holds for these values, there is a constant CCVE. □

Assume the same affine term, say $a'_1 e_1 + a_1 e_2 + b_1$ appears, up to a constant multiplicative factor, in both V^1 and V^2. This happens either when a) $c_j = a_i$, $c'_j = a'_i$ and $d_j = b_i$, or when b) $a_i = a'_j$ ($i = 1, 2$) and $b_i = b_j$.

Then, according to Theorem A.2, the line with equation $a'_1 e_1 + a_1 e_2 + b_1 = 0$ is at the same time an indifference curve and a reaction curve. The constant conjectures for players 1 and 2 are respectively, in case a): $r_1 = -a_1/a'_1 = -c_2/c'_2$ and $r_2 = -a'_1/a_1 = -c'_2/c_2$, and in case b): $r_1 = -a_1/a'_1$, $r_2 = -a'_1/a_1$. It can be checked that these values of r_1 are indeed roots of (A.12).

In order to complete the analysis, one has to investigate the second order condition (1.7), with an affine conjecture $e_j = r_j e_i + s_j$. This leads to consider functions of the form: $(Ax + B)^\mu \ (Cx + D)^\nu$. The coefficients A, B, C, D depend on the $a_i, a'_i, b_i, c_i, c'_i, d_i$ as well as on r_i and s_i. When differentiating, it follows that the sign of the second derivative is that of the second order polynomial:

$$p(x) = A^2C^2x^2(\mu + \nu)(\mu + \nu - 1) \ + \ 2ACx(\mu + \nu - 1)(AD\mu - BC\nu) \ + \\ (AD\mu + BC\nu)^2 - \mu A^2 D^2 - \nu C^2 B^2 \ .$$

The discriminant of this polynomial is $\Delta = 4(AD - BC)^2\mu\nu(\mu + \nu - 1)$.

Consequently, if $0 \le \mu + \nu < 1$, and $\mu\nu > 0$ then $p(x)$ is always negative, and the maximality condition is satisfied for any affine conjecture.

If $\mu + \nu = 1$, then $p(x)$ is actually a constant, equal to $\mu(\mu - 1)(AD - BC)^2$, hence negative. The same conclusion holds.

If $\nu = 0$ (or, symmetrically, $\mu = 0$), then we have concavity when and only when $0 \le \mu \le 1$.

In all other cases, $p(x)$ can have both signs, depending on the conjecture. A more detailed analysis is then necessary.

A.3 Polynomial consistent conjectures

Following Bresnahan (1981), we now investigate whether payoff functions that are polynomials are likely to admit consistent CVE, with conjectured reaction functions that are polynomials.

Assume therefore that the payoff functions are of the form:

$$V^i(e_i, e_j) = \sum_{(n,m)\in\mathcal{D}^i} v^i_{n,m}\, e_1^n\, e_2^m \,,$$

with $\mathcal{D}^i \subset \mathbb{N} \times \mathbb{N}$ the *finite* set of powers of monomials with a non-zero coefficient. Assume also that the conjectures $\rho_i^c(e_j)$ are polynomials, with degree P_i:

$$\rho_i^c(e_j) = \sum_{\ell=0}^{P_i} r^i_\ell\, e_j^\ell \,.$$

We shall assume that $P_i \geq 1$, $i = 1, 2$. The case of consistent Nash conjectures (zero conjectural variations) is studied specifically in Section A.4.

If the conjectured reaction functions ρ_i^c are consistent, then according to Theorem 1.3 (page 17), Equation (1.12) must hold in a neighbourhood of the candidate equilibrium. Obviously, the left-hand side of this condition is a polynomial in the variable e_j. It must therefore be identically null. But this polynomial is the sum of two terms. A necessary condition for its vanishing is that both terms have the same degree. Writing down the values of both terms, we find that:

$$V_i^i(\rho_i^c(e_j), e_j) = \sum_{\{(n,m)\in\mathcal{D}^i \;|\; n\neq 0\}} n\, v^i_{n,m} \left(\sum_{\ell=0}^{P_i} r^i_\ell\, e_j^\ell\right)^{n-1} e_j^m$$

and

$$(\rho_j^c)'(\rho_i^c(e_j))\; V_j^i(\rho_i^c(e_j), e_j) = \sum_{k=1}^{P_j} k\, r^j_k \left(\sum_{\ell=0}^{P_i} r^i_\ell\, e_j^\ell\right)^{k-1} \times \sum_{\{(n,m)\in\mathcal{D}^i | m\neq 0\}} m\, v^i_{n,m} \left(\sum_{\ell=0}^{P_i} r^i_\ell\, e_j^\ell\right)^{n} e_j^{m-1} \,.$$

From these formulas, the direct computation of the respective degrees of these polynomials in e_j yields:[2]

$$\max_{\{(n,m)\in\mathcal{D}^i \;|\; n\neq 0\}} \{(n-1)P_i + m\}$$

and

[2] We use here the fact that $P_j \neq 0$.

$$P_i\ (P_j - 1)\ +\ \max_{\{(n,m)\in\mathcal{D}^i\ \mid\ m\neq 0\}}\{nP_i + m - 1\}\ .$$

However, it may happen that the degrees are actually smaller, when the coefficients of monomials cancel. This phenomenon cannot happen if, for instance, not two distinct values of (n, m) attain the maximum in these formulas.

Assuming that no cancellation occurs for highest degree terms, and after simplifications, the equality of degrees is equivalent to:

$$\max_{\{(n,m)\in\mathcal{D}^i\ \mid\ n\neq 0\}}\{nP_j + m\}\ =\ \max_{\{(n,m)\in\mathcal{D}^i\ \mid\ m\neq 0\}}\{nP_j + m\}\ +\ P_iP_j - 1\ , \tag{A.13}$$

for $i = 1, 2$. Condition (A.13) is then a necessary condition for the conjectures to be consistent.

The case studied by Bresnahan, that is, a symmetric duopoly with linear marginal costs, corresponds to the list of powers: $\mathcal{D}^1 = \{(2,0),(1,1),(1,0),(0,0)\}$, and $P_1 = P_2 = P$. It is easily checked that the only way (A.13) can be satisfied, is that $P = 1$. The only possible conjectures are therefore affine, that is, with constant variations. When marginal costs are quadratic, then $\mathcal{D}^1$ also includes $(3,0)$. The values above for the degrees turn out to be exact: no cancellation of coefficients can occur. From condition (A.13), it is easily seen that no values of $P_i = P_j = P$ can exist. There are no polynomial consistent conjectures in this case.

A.4 Nash equilibria, Pareto optima and consistency

Part of the discussion about conjectural equilibria is summarised in the question: "is the Nash equilibrium consistent?". In order to answer this question in the present context, we have to make it more precise since a conjectural equilibrium consists in a point of equilibrium *and* variational conjectures. The question is actually if "playing Nash" by making conjectures with no variation at all ($r_i \equiv 0$ or $\rho_i^c = cst$) can be consistent. The following theorem answers this question.

Theorem A.7 *A Nash equilibrium* (e_1^N, e_2^N) *and constant conjectured reaction functions* $\rho_i^c(e_j) = e_i^N$ *(with thus a variation identically null) form a CCVE if and only if this is a Nash equilibrium in dominant strategies, or* *equivalently, if:*

$$V_i^i(e_i^N, e_j) = 0 , \qquad \forall e_j \in E_j .$$

Proof. Immediate from Theorem 1.2, page 15. Replacing $(\rho_i^c)'(\cdot)$ by 0 and $\rho_i^c(\cdot)$ by e_i^N in (1.11) gives the result. □

One appealing idea is that when a Nash equilibrium happens to be a Pareto optimum, then "playing Nash" should be somehow "consistent". In the following example, there is a Nash equilibrium which is Pareto optimal but not consistent in the sense of conjectural variations. Consider:

$$V^i(e_i, e_j) = 1 - (e_i + e_j)^2$$

and suppose that players try to maximise this function (the same for both players). It is clear that all points such that $e_i + e_j = 0$ are Nash equilibria. On the other hand, these are not consistent CVE. Indeed

$$V_i^i(e_i, e_j) = 2\,(e_i + e_j)$$

cannot not be identically zero, whatever the value given to e_i^N. The condition of Theorem A.7 for consistency is not satisfied.

Conversely, here is an example of a consistent Nash equilibrium which is not Pareto-optimal. Consider

$$V^i(e_i, e_j) = 1 - f_i(e_j)(e_i - e_i^N)^2$$

with $f_i(e_j)$ a differentiable function with $f_i(e_j^N) > 0$, and such that there exists some e_j^* with $f_i(e_j^*) < 0$. Clearly, (e_1^N, e_2^N) is a Nash equilibrium which is consistent. Indeed,

$$V_i^i(e_i, e_j) = -\,2 f_i(e_j)(e_i - e_i^N) \quad \Rightarrow \quad V_i^i(e_i^N, e_j) = 0 ,$$

so that, still from Theorem A.7, there is consistency. But this equilibrium is not Pareto-optimal since:

$$V^i(e_1^*, e_2^*) > 1 = V^i(e_1^N, e_2^N) .$$

The conclusion is that there is *a priori* no coincidence in general between Nash Equilibria, (Consistent) CVE and Pareto outcomes. The question to know whether these particular strategies can be compared, with respect to their relative values or to the payoff they provide, is addressed in Appendix B.

Appendix B

Comparison Between Conjectural Equilibria, Nash Equilibria and Pareto-Efficient Outcomes

Our aim in this annex is to compare the CVE and the Nash Equilibrium (NE), and thus to appraise the impact on agents' behaviours of the underlying intertemporal forces that yield the conjectures. From the existing results, it is not clear how one can know, in advance, the consequences of non-zero conjectures on behaviours. Our aim is: *i*) to identify situations where it is indeed possible, ***a priori***, to know which kind of noncooperative concept Pareto dominates the other, *ii*) to provide out the corresponding theoretical explanations.

To do so we focus on a static symmetric game which is intended to capture long run interactions between identical agents, and on symmetric solutions.

The economic situations at hand can be divided into two families, depending on whether they admit a stable Nash equilibrium and an interior Pareto solution (family 1) or not (family 2). Within each family, the sign of the externalities (positive or negative effect of the rival actions on a player's payoffs) together with the properties of conjectures (their sign and their absolute values) not only indicates how to rank the action levels associated with the NE and the CVE, but also allows one to predict which kind of behaviour leads the players to the most favourable outcome. It turns out that the qualitative results prevailing for family 1 are reversed for family 2. This classification is useful in that outcomes and payoffs need not be calculated to assess the impact of conjectures on players' payoffs; the only relevant pieces of information are the sign of second order derivatives of the payoff function and the properties of conjectures, *i.e.* the description of the game. We then study in which kind of game reasonable conjectures, *i.e.* consistent conjectures, belong to the set of conjectures that produces superior outcomes.

B.1 Two-player games

For the purposes of this chapter, we shall consider symmetric games, where the strategy space of each player is a compact convex subset of $\mathbb{R}$:

$$E_1 = E_2 = E = [\underline{E}, \overline{E}] \ ,$$

Player i's payoff function is twice continuously differentiable: $V^i \in C^2(E \times E)$, $i = 1, 2$. By convention, the first variable of a player's payoff function is the strategy of this player. We consider symmetric payoffs, *i.e.*:

$$V^2(e_2, e_1) = V^1(e_2, e_1) = V(e_2, e_1) \ .$$

As indicated by the last equality only one function needs to be considered, the first argument being sufficient to indicate the player associated with the payoff. For instance, in a symmetric duopoly model (see Section C.1), the payoff functions are:

$$V^1(e_1, e_2) = (a - b(e_1 + e_2))e_1 - ce_1 \ ,$$
$$V^2(e_2, e_1) = (a - b(e_2 + e_1))e_2 - ce_2 \ .$$

Clearly, only one function $V(x, y)$ is needed to describe the payoffs:

$$V(x, y) = (a - b(x + y))x - cx \ .$$

In order to assess the impact of conjectures on decisions it is necessary to compare three possible scenarios: *i*) agents behave noncooperatively taking rival decisions as given, *ii*) agents decide noncooperatively given some conjectures on rival strategies, *iii*) decisions are taken by a benevolent authority who is in charge with the interest of both agents. Relating the outcomes of these different scenarios we can learn how conjectures distort the individual decisions and what are the consequences in terms of efficiency.

We shall work throughout this section with *interior symmetric* solutions only. Clearly, this choice discards presumably interesting cases; it however fits well with a large body of the literature where, to our knowledge, outcomes turn out to be interior.

In the first scenario, denote by (e^N, e^N) a symmetric Nash equilibrium. A necessary condition satisfied by any interior symmetric Nash equilibrium is:

$$V_1(e^N, e^N) \ = \ 0 \ . \tag{B.1}$$

Turning to the second scenario, players have conjectures about the rival behaviour. In what follows, we shall adopt *constant* conjectures, and therefore set $r(e) = r$, but the theorems B.1 and B.2 to follow have straightforward extensions to non-constant conjectures.

Let us denote (e^c, e^c) a symmetric conjectural variations equilibrium. A necessary condition satisfied by any interior symmetric CVE is, according to Theorem 1.1 page 12:

$$V_1(e^c, e^c) \;+\; r\, V_2(e^c, e^c) \;=\; 0\,. \tag{B.2}$$

In the last scenario, the pair (e^P, e^P) denotes a symmetric Pareto optimum. In this symmetric context, it makes sense to focus on the utilitarian criterion to derive a Paretian outcome. In other words the Pareto optimum under consideration will be the one that optimises the sum of the payoffs $V(e_1, e_2) + V(e_2, e_1)$. A necessary condition satisfied by any interior symmetric Pareto optimum for this utilitarian criterion is:

$$V_1(e^P, e^P) \;+\; V_2(e^P, e^P) \;=\; 0\,. \tag{B.3}$$

Observe that (B.3) is a particular case of a conjectural equilibrium, with $r = 1$, whereas (B.1) results for $r = 0$.

The following definitions will be useful in the discussion to follow. Let:

- $G^N(e) = V_1(e, e)$ stands for the marginal effect of e_i on agent i's payoff, evaluated at a symmetric outcome,
- $G^c(e) = V_1(e, e) + rV_2(e, e)$ be the marginal effect of e_i on agent i's payoff given a conjectured relation $(\rho_j^c)'(e_i) = r$ between strategies, still for a symmetric outcome,
- $G^P(e) = V_1(e, e) + V_2(e, e)$ be the marginal effect of e_i on the utilitarian criterion, evaluated at symmetry,
- and let $W(e) = V(e, e)$.

We have:

i) $G^c(e) - G^N(e) = V_2(e, e)r$.
ii) $G^P(e) - G^c(e) = V_2(e, e)(1 - r)$.
iii) $G^P(e) - G^N(e) = V_2(e, e)$
iv) $W'(e) = G^P(e)$. Therefore, any solution (e^P, e^P) of (B.3) which is a global maximum is such that

$$W(e^P) \;=\; V(e^P, e^P) \;\geq\; V(e, e) \;=\; W(e),\ \forall e \in E\,.$$

B.1.1 *Main results*

As we now proceed to show, the impact of conjectures on the outcome of the game is heavily driven by the sign of the function:

$$V_{11}(e,e) + V_{12}(e,e) \ ,$$

as well as the sign of the externality V_2, and the position of the conjectural variation r with respect to the values 0 and 1.

Technically the importance of some of the above elements can be grasped easily. The conjectural variations equilibrium e^c and the corresponding payoffs are configured by the constant conjecture r,

$$e^c = e^c(r) \quad \text{and} \quad \widetilde{V}(r) = V(e^c(r), e^c(r)).$$

A small change in the conjecture r will have marginal impacts on equilibrium $(e^c)'(r)$ and payoffs as well:

$$\widetilde{V}'(r) = \big(V_1(e^c(r), e^c(r)) + V_2(e^c(r), e^c(r))\big)\ (e^c)'(r)\ .$$

From (B.2), one obtains by differentiating with respect to r:

$$(e^c)'(r) \ = \ -\frac{V_2}{(V_{11}+V_{12}) \ + \ r(V_{12}+V_{22})}(e^c, e^c)$$

so,

$$(e^c)'(0) \ = \ -\frac{V_2}{V_{11}+V_{12}}(e^N, e^N)\ ,$$

which makes clear that locally, starting from a Nash (zero) conjecture, the sign of $(e^c)'$, hence the sign of $\widetilde{V}'$, depends on the sign of externalities and the sign of $V_{11} + V_{12}$. Note also that $V_1(e^c(r), e^c(r)) + V_2(e^c(r), e^c(r))$ is $G^P(e^c)$, so that the sign of $\widetilde{V}'(r)$ is also related to the location of e^c with respect to the zero of function $G^P(e)$.

With these technical informations in mind, consider the following assumptions:

(H1) There exists a unique interior Nash equilibrium in E^2 and

$$V_{11}(e,e) + V_{12}(e,e) \le 0 \qquad \forall e \in E. \tag{B.4}$$

Alternatively the reverse inequality will also prove useful:

(H1') There exists a unique interior Nash equilibrium in E^2 and

$$V_{11}(e,e) + V_{12}(e,e) \geq 0 \qquad \forall e \in E. \tag{B.5}$$

(H2) There exists at most one zero of the function G^P in the interior of E.

For the moment, we shall merely note that property (B.4) implies that $G^N(e)$ is a decreasing function; on the contrary (B.5) implies that $G^N(e)$ is an increasing function.

Moreover, if there exists a unique Nash equilibrium in a symmetric game, this equilibrium must be symmetric.[3] A consequence of (**H1**) or (**H1'**) is therefore the existence of a unique $e^N \in E$ such that $G^N(e^N) = 0$. We shall come back to these assumptions in Section B.1.2.

Two main general theorems are developed. They compare Nash, conjectural variation and Pareto strategies and also the corresponding payoffs. The results of the comparisons depend on the sign of externalities and conjectures. These Theorems B.1 and B.2 are associated respectively with (B.4) and (B.5). We then proceed to a discussion of the assumptions and the results.

Theorem B.1 *Assume (**H1**) and (**H2**) are satisfied. Then for the (symmetric) Nash equilibrium, any symmetric conjectural variations equilibrium and the (symmetric) Pareto optimum (if it exists):*
Case 1 *If $V_2 > 0$ (positive externality) and $r > 0$ (positive conjectural variation), then:*
either $r \leq 1$, and then

$$e^N \leq e^c \leq e^P \tag{B.6}$$

and

$$V(e^N, e^N) \leq V(e^c, e^c) \; ; \tag{B.7}$$

or $r > 1$, and then:

$$e^N \leq e^P \leq e^c \tag{B.8}$$

and it is not possible to order the payoffs associated with the outcomes.
Case 2 *If $V_2 < 0$ (negative externality) and $r > 0$ (positive conjectural*

[3] For if (e_1, e_2) is a Nash equilibrium, then so is (e_2, e_1) from the symmetry assumptions on V^1 and V^2. If $e_1 \neq e_2$, this makes two distinct Nash equilibria, a contradiction.

variation), then:
either $r \leq 1$, and then

$$e^P \leq e^c \leq e^N$$
$$V(e^N, e^N) \leq V(e^c, e^c) ;$$

or $r > 1$, and then $e^c \leq e^P \leq e^N$, and it is not possible to order the payoffs associated with the outcomes.

Case 3 *If $V_2 > 0$ and $r < 0$ then*

$$e^c \leq e^N \leq e^P$$
$$V(e^c, e^c) \leq V(e^N, e^N) .$$

Case 4 *If $V_2 < 0$ and $r < 0$, then*

$$e^P \leq e^N \leq e^c$$
$$V(e^c, e^c) \leq V(e^N, e^N) .$$

Proof. The results simply follow from the adequate ordering between the curves G^N, G^P and G^c, which implies the ordering between their zeros. Case 1 and Case 4 are illustrated in Figure B.1.

Case 1: The assumptions of positive externalities and positive conjectures imply $G^c(e) - G^N(e) = rV_2(e, e) > 0$.
If $r \leq 1$ then $G^P(e) - G^c(e) \geq 0$ and since $r > 0$, then $G^c(e) - G^N(e) > 0$. Therefore:

$$G^N(e) < G^c(e) \leq G^P(e) \qquad \forall e \in E . \tag{B.9}$$

This inequality, the assumption that $G^N(e)$ is a decreasing function, and (**H2**) clearly imply that (B.6) holds for any conjectural equilibrium e^c.

If there exists e^{p_1} solution of $G^P(e) = 0$ in E, then $G^P(e) \geq 0$ if $e \leq e^{p_1}$. If $G^P(e) \neq 0$ in E, then $G^P(e) > 0$ in E. In the latter case $G^P(e)$ can not be negative because it is in contradiction with (B.9). This implies that $W(e)$ is an increasing function for all $e \leq e^{p_1}$ or for all $e \in E$, and because $e^N \leq e^c$, $V(e^N, e^N) \leq V(e^c, e^c)$, so (B.7) holds.

If $r \geq 1$ then $G^P(e) - G^N(e) = V_2(e) > 0$ and $G^P(e) - G^c(e) < 0$, then

$$G^N(e) < G^P(e) < G^c(e) \qquad \forall e \in E$$

so (B.8) holds. The fact that no generic comparison is possible between $V(e^N, e^N)$ and $V(e^c, e^c)$ stems from the fact that the maximum of the function $V(e,e)$ occurs between e^N and e^c. This is illustrated in Figure B.2. A concrete example of this is detailed in Section C.2.

The proofs of the three other cases are similar and left to the reader.□

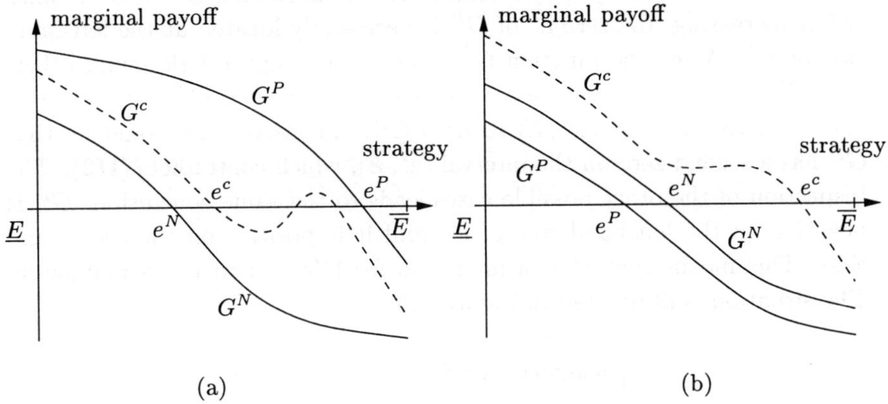

Fig. B. 1 Illustrations of the proof of Theorem B.1, Case 1 (a) and Case 4 (b).

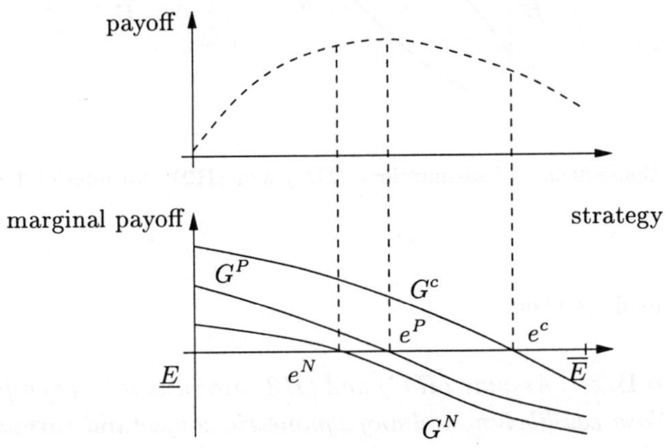

Fig. B.2 The case where $e^N < e^P < e^c$: no general comparison of payoffs is possible.

The next theorem makes use of assumptions (**H1'**) and (**H2**). Taken simultaneously these assumptions single out a situation where there exists no interior Pareto solution. Before turning to the results, it is worth understanding what are the implications of the fact that G^N is an increasing function on E (a consequence of (**H1'**)) and that there exists at most one zero for the function G^P on E (**H2**). Take for instance a situation of positive externalities and positive conjectures: in such case $G^P > G^N$. Since G^N is increasing, the zero e^* of G^P is necessarily located at the left-hand side of e^N. At e^* the function G^P is necessarily upward slopping; otherwise G^P becomes negative between e^* and e^N, and since G^N takes positive values on the interval $(e^N, \overline{E}]$ so does G^P: a consequence would be that G^P has a second zero on the interval $(e^*, \overline{E}]$, which contradicts (**H2**). The inspection of the other possible cases leads to the same conclusion: G^P is negative on the left-hand side of e^*, and it is positive on the right-hand side. This means that e^* is a minimum for $V(e,e)$ and not a maximum. The situation is illustrated in Figure B.3.

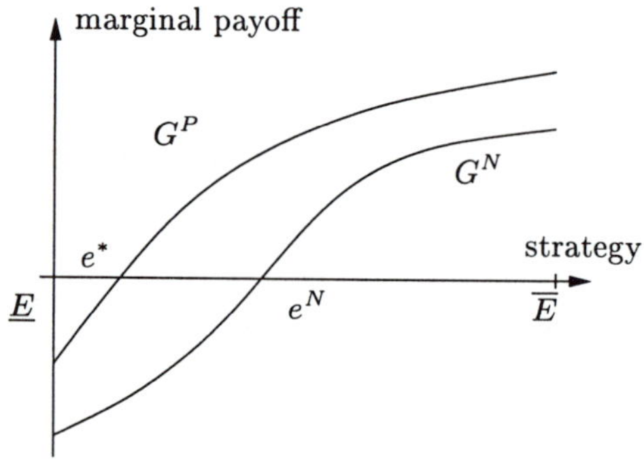

Fig. B.3 Consequence of assumptions (**H1'**) and (**H2**): no interior Pareto outcome exists.

The result is then:

Theorem B.2 *Assume (**H1'**) and (**H2**) are satisfied. Then for the (symmetric) Nash equilibrium and any symmetric conjectural variations equilibrium, we have:*

Case 1 *If $V_2 > 0$ (positive externality) and $r > 0$ (positive conjectural variation): then*

$$e^N \geq e^c \tag{B.10}$$

and if $r \leq 1$ *then*

$$V(e^N, e^N) \geq V(e^c, e^c) . \tag{B.11}$$

If $r > 1$, *it is not possible to order the payoffs associated with the outcomes.*
Case 2 *If* $V_2 < 0$ *(negative externality) and* $r > 0$ *(positive conjectural variation), then*

$$e^c \geq e^N ,$$

and if $r \leq 1$ *then*

$$V(e^N, e^N) \geq V(e^c, e^c) .$$

If $r > 1$, *it is not possible to order the payoffs associated with the outcomes.*
Case 3 *If* $V_2 > 0$ *and* $r < 0$ *then*

$$e^c \geq e^N$$

and

$$V(e^c, e^c) \geq V(e^N, e^N) .$$

Case 4 *If* $V_2 < 0$ *and* $r < 0$, *then*

$$e^c \leq e^N$$

and

$$V(e^c, e^c) \geq V(e^N, e^N) .$$

Proof. **Case 1:** From the assumptions of positive externalities and positive conjecture we have that $G^c(e) - G^N(e) = rV_2(e, e) > 0$.

This inequality implies that $e^c < e^N$. What can we learn about the payoffs? As observed above, $G^P > G^N$, has an increasing slope at the point e^* where $G^P(e^*) = 0$, is negative at the left-hand side of this point and positive at the right-hand side. Thus the unique zero e^* of G^P is a minimum of $W(e)$, which means that the Pareto solution is not interior. If $r \leq 1$ then $G^P(e) - G^c(e) \geq 0$; e^c and e^N are located on the right-hand side of e^*, *i.e.* on the segment where an increase of e induces an increase of the players' payoffs: $V(e^c, e^c) \leq V(e^N, e^N)$. If $r > 1$, then $G^P(e) - G^c(e) < 0$. Therefore e^* lies between e^c and e^N: it is not possible to order the payoffs. The proofs of the three other cases are similar. □

B.1.2 *Discussion*

Intuitions underlying Theorems B.1 and B.2

When a typical agent forms conjectures, she no longer considers rival decisions as externalities. She behaves as if there were an existing functional relation between her actions and the externalities. Let us consider successively some cases of Theorems B.1 and B.2. First take case 2 of Theorem B.1, negative externalities and positive conjectures, and remind the first order condition attached to the symmetric conjectural equilibrium:

$$V_1(e^c, e^c) \;=\; -\,r\; V_2(e^c, e^c)\;. \tag{B.12}$$

With $r > 0$, when agent i modifies her action she exerts a direct marginal effect on her payoff, $V_1(.,.)$, and a conjectured indirect marginal effect on the negative externality $-rV_2(.,.)$. This "externality" can be interpreted here as a cost induced by the action e_i. Since the externality is negative, the analogy with a cost naturally arises. As usual, the optimal choice e^c, evaluated at a symmetric outcome, is the one that realises the equality between the marginal benefit $V_1(e,e)$ and the marginal cost $-rV_2(e,e)$. An alternative interpretation would be to say that agent i bears a ratio of the externality she exerts to the second player, that is to say she internalises r percent of the externality she generates.

This is better illustrated by a drawing of the functions $V_1(e,e)$ and $-rV_2(e,e)$. In Figure B.4 (top), the decreasing curve represents the direct marginal effect of agent i's strategy computed at a symmetric outcome; the increasing curves represent the conjectured indirect marginal effects of i's action for three different value for the conjecture r, more precisely $r = 0.5$, 0.75 and 0.99. As r increases the upward slopping curves switch towards the upper left corner. The symmetric conjectural equilibrium decision e^c is the point on the x-axis for which the increasing and the decreasing curves intersect. The symmetric Nash decision, that results for $r = 0$, corresponds to the point e^N such that $V_1(e^N, e^N) = 0$. The higher the conjecture r, the lower e^c.

When $0 < r < 1$, agent i partially internalises the externality; she underestimates the negative effect of her decision on the social interest. However as r increases towards unity the inefficiency of noncooperative decisions vanishes. Social effects of decisions are perfectly taken into account when $r = 1$, and Pareto efficiency obtains. When $r > 1$, agent i over-estimates the costs she produces on the other player. Consequently as r increases from unity, the loss of efficiency increases and eventually may end up in a

Pareto deterioration as compared to the Nash equilibrium.

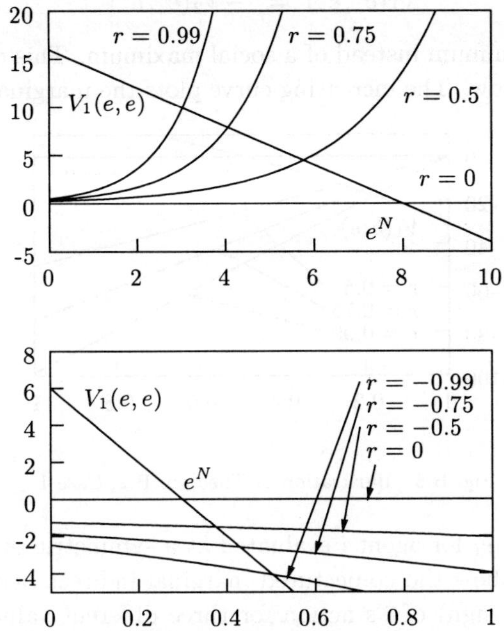

Fig. B.4 Illustration of Theorem B.1, Case 2 (top) and Case 4 (bottom).

What happens if the conjecture is negative (and externalities are still negative: case 4 of Theorem B.1)? The right-hand side of (B.12) changes its sign. It is as if the agents make a mistake and consider the negative externality as an additional benefit instead of a cost: negative conjectures modify behaviours in the wrong direction. This case is depicted in Figure B.4 (bottom). The three negative curves correspond to the conjectured indirect marginal effects of i's action for three different negative values for the conjecture r, $r = -0.5$, -0.75 and -0.99. As $|r|$ increases the curves switch downwards and the conjectural equilibrium decision e^c increases, leaving the agents worse-off (remember that the direction of Pareto amelioration is the one where e^c decreases). Cases 1 and 3 can be discussed in much the same way as cases 2 and 4.

It remains to deliver the intuition for Theorem B.2. Consider a case of positive externalities and positive conjectures (case 2 of Theorem B.2). Remember that under (**H1'**) we are dealing with situations where the decision

e^* such that:

$$V_1(e^*, e^*) = -V_2(e^*, e^*)$$

gives a social minimum instead of a social maximum. This case is depicted in Figure B.5 below. The increasing curve plots the marginal direct benefit

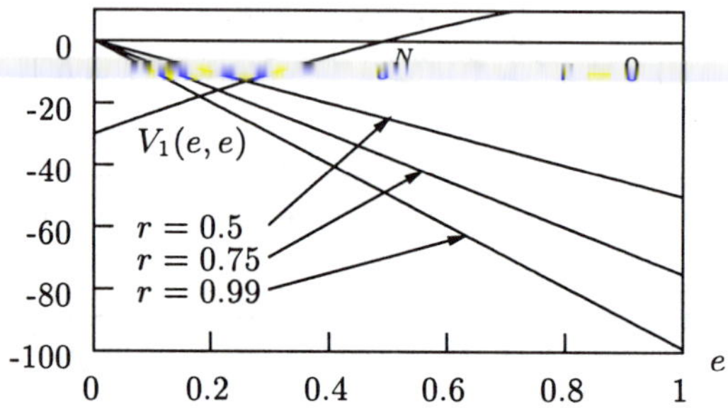

Fig. B.5 Illustration of Theorem B.2, Case 1.

$V_1(e, e)$ of action e_i for agent i evaluated at a symmetric outcome; the decreasing curves show the conjectured marginal indirect benefit $-rV_2(e, e)$ (with a negative sign) of i's action for three different values for the conjecture r (as before $r = 0.5$, 0.75 and 0.99). As r increases the decreasing curves switch towards the lower left corner and the conjectural equilibrium decision e^c decreases and tends to, for $r = 1$, a social minimum, causing a Pareto-deterioration.

Assumptions on the payoff functions

A number of remarks are in order about (**H1**), (**H1'**) and (**H2**). In this discussion we shall assume that $V_{11} < 0$ in $E \times E$ (*i.e.* the concavity of V as a function of e_1), which is natural in most economic situations.

First, observe that conditions (B.4) and (B.5) are related to the slope of the Nash best responses of players. Indeed, player 1's best response given implicitly by $V_1(e_1, e_2) = 0$ satisfies

$$(\chi_1^N)'(e_2) = -\frac{V_{12}(\chi_1^N(e_2), e_2)}{V_{11}(\chi_1^N(e_2), e_2)}$$

in $E \times E$. If (B.4) holds, and with $V_{11} < 0$, then $-V_{12}/V_{11} \leq 1$. Conversely, if (B.5) holds, then $-V_{12}/V_{11} \geq 1$. In other words, (**H1**) is equivalent to requiring that there exists a unique interior Nash equilibrium, and that the slope of the Nash best responses is smaller than 1. Similarly (**H1'**) amounts to requiring that there exists a unique interior Nash equilibrium, and that the slope of the Nash best response is larger than 1. An implication of the above remarks is that under (**H1'**) the Nash equilibrium is unstable, a quite undesirable property. Under (**H1**), if the game features strategic complementarity ($V_{12} > 0$), *i.e.* when the Nash best responses are upward sloping, then the Nash equilibrium under consideration is stable. When the game otherwise features strategic substitutability ($V_{12} < 0$), in other words when the Nash best responses are downward sloping, the Nash equilibrium under (**H1**) is stable if and only if $V_{11} < V_{12}$.

A sufficient condition to have (**H2**) is to require $G^P(e)$ to be a decreasing function in E, or equivalently, that $W(e) = V(e,e)$ be a concave function. Observe that $W(e) = V(e,e)$ is a concave function for symmetric oligopoly with decreasing and concave inverse demand and convex costs, and also for models of contribution to public goods (see Section 1.4). We do not mean however that the payoff function corresponding to those games are concave; only the function of one variable obtained when those payoff functions are evaluated along a symmetric outcome is concave. If $V(e_1, e_2)$ is a concave function in $E \times E$ (as a function of the two variables) then $G^N(e)$, $G^c(e)$ (when r is constant) and $G^P(e)$ are decreasing functions in E. But unfortunately this is not generally the case in economic applications. Finally one could ask whether the inequalities of assumptions (**H1**) and/or (**H1'**) have some bearings with those underlying the classes of supermodular and submodular games. Supermodularity requires that the Nash best response functions χ_i^N be decreasing. We instead have a requirement on the slope of the function $G^N(e)$, which is different. Expressed in terms of second-order derivatives, concavity in each variable and supermodularity would imply $V_{11} \leq 0$ and $V_{12} \geq 0$, therefore $(G^N)'(e) = V_{11} + V_{12}$ can have any sign. On the other hand, submodularity and concavity do imply $\left(G^N\right)'(e) \leq 0$ of assumption (**H1**); but clearly the inequality of (**H1**) does not necessarily imply submodularity.

Finally, our theorems apply to cases with an unique *interior* Nash equilibrium. Since the strategy space is a closed set, one could ask about the possibility to have a Nash equilibrium at the boundary. It is easily proved, using Kuhn-Tucker conditions, that only when (**H1'**) is verified a Nash equilibrium at the boundary can exist. We discuss in Section C.4 a case where this indeed happens, and we show how the results of Theorem B.2 can be completed in such a situation.

Implication of the results

Concerning the theorems, note that the sign of the externalities is irrelevant to compare the payoffs corresponding to scenarios i) and ii). Indeed, from Theorem B.1, if (**H1**) and (**H2**) hold, then whatever the sign of externalities:

$$\text{if}: 0 < r \leq 1 \text{ then } V(e^N, e^N) \leq V(e^c, e^c)$$

$$\text{if}: r < 0 \qquad \text{then } V(e^c, e^c) \leq V(e^N, e^N) \ .$$

In other words:

> When the slopes of the Nash best response functions are less than unity, positive but less than unity conjectural variations lead to a Pareto-improvement. Negative conjectural variations lead to a Pareto-worsening.

On the other hand, if (**H1'**) and (**H2**) hold, *i.e.* if the Nash equilibrium is unstable and there is no interior Pareto solution, the results are reversed; whatever the sign of externalities:

$$\text{if}: 0 < r \leq 1, \text{ then } V(e^N, e^N) \geq V(e^c, e^c)$$

$$\text{if}: r < 0, \qquad \text{then } V(e^c, e^c) \geq V(e^N, e^N).$$

In other words:

> When the slopes of the Nash best response functions are larger than unity, positive but less than unity conjectural variations lead to a Pareto-worsening. Negative conjectural variations lead to a Pareto-improvement.

Finally, note that assumption (**H2**) is essential in order to obtain a ranking of Pareto outcomes and conjectural equilibria. For instance, in Theorem B.1, Case 1, without this property, it could be that $e^P \leq e^c$ for some Pareto optimum e^P and some CVE e^c and comparing the payoffs would not be possible.

B.2 Many players games

As in Section 1.3.9, we consider the class of games where for each player, the contributions of all other players to the externality are aggregated. The payoff functions under consideration are $V(e_i, y_i)$ where $y_i = \sum_{j \neq i} e_j$. The payoff of agent i thus depends on her strategies and the sum of the rival strategies. Each agent conjectures how the other players, as a whole, react to her choices. It is assumed that these conjectures are all identical. *i.e.* $\frac{\mathrm{d}y_i}{\mathrm{d}e_i} \equiv r$, $\forall i$. Given the form of the payoff functions, marginal external effects read as:

$$V_{e_j} = V_{y_i}$$

for all $j \neq i$. Here, the notations V_e and V_y still denote partial derivatives of the function V. An interior symmetric Nash equilibrium $(e^N, \ldots, e^N)$ verifies:

$$V_e(e^N, (n-1)e^N) = 0 \ .$$

An interior symmetric conjectural variations equilibrium $(e^c, \ldots, e^c)$ satisfies:

$$V_e(e^c, (n-1)e^c) \ + \ r \ V_y(e^c, (n-1)e^c) \ = \ 0 \ .$$

Finally, an interior symmetric Pareto outcome $(e^P, \ldots, e^P)$ must verify:

$$V_e(e^P, (n-1)e^P) + (n-1)V_y(e^P, (n-1)e^P) = 0 \ .$$

It is straightforward to extend the study and results established for two players to this symmetric n-player framework. Indeed, modify the definitions of functions $G^N(e)$, $G^c(e)$ and $G^P(e)$ of Section B.1, by replacing "e" with "$(n-1)e$" as second argument of V_1 and V_2. Then, e^N, e^c and e^P are solution of $G^N(e^N) = 0$, $G^c(e^c) = 0$ and $G^P(e^P) = 0$, respectively. The differences between the three functions have the same form as before, with V_2 replaced by V_y and $(1-r)$ replaced by $(n-1-r)$. But the sign of the differences $G^N - G^c$, $G^N - G^P$, $G^P - G^c$ are not altered. Assumption (**H2**) does not need to be changed. In order to state the new (**H1**) one simply has to replace the assumption

$$V_{11}(e,e) + V_{12}(e,e) \ \leq \ 0$$

by

$$V_{ee}(e, (n-1)e) + (n-1)V_{ey}(e, (n-1)e) \ \leq \ 0 \ .$$

To obtain the new (**H1'**), simply reverse this above inequality. Then Theorems B.1 and B.2 hold with "$r \gtreqless 1$" replaced by "$r \gtreqless n-1$".

B.3 Consistent conjectures

Among all conjectural equilibria, those which are consistent deserve a separate treatment. In order to apply Theorems B.1 and B.2 to such endogenous conjectures, it is necessary to know their sign. In the present symmetric situation, Corollary 1.5 (page 18) gives the answer: two conjectures r_+ and r_-, at most, are possible, and they have the same sign, which is that of:

$$-\frac{V_{11}+V_{22}}{V_{12}}.$$

In several models, the payoff function has a quadratic form: this is the case in Cournot's duopoly model and the model of competition between regions, reviewed respectively in Sections C.1 and C.3. In such cases, the V_{ij} are constant parameters and in general $V_{11} < 0$ and $V_{22} \leq 0$ to ensure the existence of a Nash equilibrium. Therefore, *the sign of every consistent conjecture r is given by the sign of V_{12}*. For example if $V_{12} < 0$ then consistent conjectural variations are negative. In addition, $V_{11} + V_{12} < 0$ so Theorem B.1 case 1 or 2 apply (provided, of course, that the other assumptions of Theorem B.1 are satisfied and a consistent CVE exists).

Appendix C

Examples and Illustrations

This appendix reviews the principal examples of economic models from the literature, where (static) Conjectural Variations Equilibria have been studied. Most of these examples are used as illustrations in other parts of this monograph, be it only in a simplified version.

When describing these economic models, we go through the following steps. First, we discuss the application of the qualitative comparison results of Appendix B (Theorems B.1 and B.2), under the most general assumptions. Then, we concentrate on the computation of (consistent) equilibria with *constant* conjectures. We perform the detailed computations, and when appropriate, show that these examples are particular cases of the general results of Appendix A (Theorem A.2 or Theorem A.4, for instance).

In the first model, we revisit the duopoly with quantity as the choice variable. The second example deals with the model of voluntary contributions of a public good. The third one is a model of competitions between regions. Finally, we discuss a model of aggregate-demand externalities.

C.1 Cournot's duopoly

In Chapter 1, Section 1.4.1, we have shown that in a Cournot's duopoly with a constant marginal cost, and for any (reasonable) inverse demand function, there is a constant CCVE with variation -1. We begin by revisiting this example in a general setting, in order to apply the qualitative results of Section B.1.1. We then concentrate on the case of a linear inverse demand function and a linear cost function, in order to state quantitative results.

Consider the general setting of Cournot's duopoly with profit function:

$$V^i(e_i, e_j) \;=\; p(e_i + e_j)e_i \;-\; C(e_i)\;.$$

Here, $p(\cdot)$ stands for the inverse demand function. It satisfies $p'(\cdot) \leq 0$ (the higher the quantities, the lower the price). The cost function C is increasing and convex. The question of existence and uniqueness of a Cournot-Nash equilibrium has been studied extensively in the literature. See for instance Tirole (1988) for a survey of those results.[4] The two following assumptions to ensure existence of a Nash Equilibrium are borrowed from this survey: $p''(\cdot) < 0$ (strict concavity of the inverse demand function) and $p(0) > C'(0)$. To these conditions, we add the possibility that $p(\cdot)$ be a strictly decreasing affine function.

Let us check whether the results of Appendix B may apply. Inequality (B.4) in (**H1**) translates here as:

$$2p''(\cdot)e_i + 3p'(\cdot) - C''(\cdot) \ \leq \ 0 \ .$$

It is automatically met under the above assumptions. Assumption (**H2**) involves the consideration of the function:

$$G^P(e) \ = \ 2p'(2e)e \ + \ p(2e) \ - \ C'(e) \ .$$

As we have restricted attention to strictly concave (or strictly decreasing affine) inverse demand functions, it follows that G^P is monotonously decreasing, so (**H2**) holds. Finally, (**H1**) and (**H2**) are satisfied. Consider now externalities. We have $V^i_j(e_i, e_j) = e_i p' < 0$ as $p' < 0$. Consequently, Theorem B.1, Cases 2 or 4, applies. It is therefore possible to compare *a priori* the outcomes and the profit values of the Nash equilibrium, the CVE and Pareto outcomes, provided the two last ones exist.

We now turn to the particular case of a linear inverse demand function, and a linear cost function. The profit function has the form:

$$V^i(e_i, e_j) \ = \ (a - b(e_i + e_j))e_i \ - \ ce_i \ .$$

The parameters a, b and c are strictly positive. When $a > c$ (that is, the condition $p(0) > C'(0)$ assumed above), (**H1**) and (**H2**) are satisfied and the externality is negative.
Moreover, $V^i_{ii} = -2b < 0$, $V^i_{ij} = -b < 0$, and $V^i_{jj} = 0$. Accordingly, by Corollary 1.5 on page 18 (see also Section B.3), the sign of the consistent conjectures must be negative.

When discussing this case in Chapter 1, Section 1.4.1, we have found that the CVE with a constant conjecture r exists if $r > -1$, and is given

[4]More precisely, the reader is invited to consider the supplementary section 5.7 of Tirole's book.

by:

$$e^c = \frac{a-c}{b(3+r)} .$$

It is easy to check directly that Theorem B.1, Case 4 holds since for any $-1 < r < 0$,

$$e^P = \frac{a-c}{4b} < e^N = \frac{a-c}{3b} < e^c = \frac{a-c}{(3+r)b} ,$$

and that the CVE yields a lower profit than does the Nash equilibrium.

We proceed to show that the existence of a CCVE is a consequence of the results of Section A.2. Indeed, the profit function is here of the form:

$$\begin{aligned} V^i(e_i, e_j) &= e_i \, p(e_i + e_j) \; - \; ce_i \\ &= e_i(p(e_i + e_j) - c) . \end{aligned} \tag{C.1}$$

Let us call e_0 the quantity such that $p(e_0) = c$. Such a value exists and is unique under the assumptions made above on the function $p(\cdot)$. Then (C.1) is of the form (A.5) with

$$\rho_j^c(e_i) = e_0 - e_i \ , \quad g_i(e_j) = 1 \ , \quad h_i(e_i) = e_i \ ,$$
$$\varphi_i(u, v) = p(u + e_0) - c \ , \quad F^i(e) = e \ .$$

We are therefore in the situation of Theorem A.4, *iii)* since $\varphi(0,0) = p(e_0) - c = 0$, and $\rho_i^c \circ \rho_j^c(e_i) = \rho_i^c(e_0 - e_i) = e_0 - (e_0 - e_i) = e_i$ for all e_i. Theorem A.2 also applies (see the remark after the proof of Theorem A.4 on page 126), and all points of the line $e_1 + e_2 = e_0$ are, together with the variational conjectures $r_j(e_i) = -1$, consistent equilibria. This line is an iso-profit curve for both firms, for the profit 0. Of course, as concluded by Bresnahan (1981), this situation is not desirable for the firms.

C.2 Voluntary contributions to a public good

In this section, we review results on the model of voluntary contributions to a public good as studied in Itaya and Dasgupta (1995), and introduced in Chapter 1, Section 1.4.3. The utility function are of the Cobb-Douglas form:

$$V^i(e_i, e_j) \; = \; (I_i - e_i)^{\alpha_i}(e_i + e_j)^{1-\alpha_i} \tag{C.2}$$

for $i = 1, 2$, with the condition $\alpha_1 + \alpha_2 < 1$.

First, we restrict our attention to the symmetric case and discuss qualitative results. The utility function is here:

$$V^i(e_i, e_j) \;=\; (I - e_i)^\alpha (e_i + e_j)^{1-\alpha} \;,$$

where $0 < \alpha < 1/2$.

The first derivative writes as:

$$V_i^i(e_i, e_j) \;=\; ((1-\alpha)(I - e_i) - \alpha(e_i + e_j))\,(I - e_i)^{\alpha-1}(e_i + e_j)^{-\alpha} \;.$$

The externality is:

$$V_j^i(e_i, e_j) = (1-\alpha)(I - e_i)^\alpha (e_i + e_j)^{-\alpha} \;>\; 0 \;.$$

Then, we have:

$$\begin{aligned} V_{ii}^i + V_{ij}^i &= -\alpha(1-\alpha)(I - e_i)^{\alpha-2}(e_i + e_j)^{1-\alpha} \\ &\quad -3\alpha(1-\alpha)(I - e_i)^{\alpha-1}(e_i + e_j)^{-\alpha} \\ &\quad -2\alpha(1-\alpha)(I - e_i)^{\alpha}(e_i + e_j)^{-1-\alpha} \;<\; 0 \;. \end{aligned}$$

Hypothesis (**H1**) is therefore satisfied. Considering now symmetric Pareto outcomes and Hypothesis (**H2**), we have:

$$\begin{aligned} G^P(e) &= 2I\;(I - e)^{\alpha-1}\;(2e)^{-\alpha}\;((1-\alpha)I - e) \\ (G^P)'(e) &= -\alpha(1-\alpha)\;I^2\;(I - e)^{\alpha-2}\;(2\,e)^{-1-\alpha} \;<\; 0 \;. \end{aligned}$$

Hypothesis (**H2**) is therefore satisfied. The Pareto outcome is $e^P = (1-\alpha)I$. Since $V_j^i > 0$, this is the situation of Theorem B.1, Case 1 or Case 3.

What about consistent conjectural equilibria? Itaya and Dasgupta have shown the existence of a constant CCVE given by (see below in this section):

$$r \;=\; -\frac{\alpha}{1-\alpha} \;<\; 0 \;, \qquad e^c \;=\; (1 - 2\alpha)I \;.$$

We are therefore in Case 3 of Theorem B.1, and the consistent conjecture yields lower payoffs than the Nash equilibrium.

Finally, we come back to the non-symmetric form (C.2) and discuss the existence of a constant CCVE. This example verifies the general form of Theorem A.6 with $a_i' = -1$, $a_i = 0$, $b_i = I_i$, $\mu_i = \alpha_i$, $c_i' = c_i = 1$, $d_i = 0$, and $\nu_i = 1 - \alpha_i$.

The constant CCVE, if they exist, are the solutions of Equation (A.12). Given the values above, this equation reduces to:

$$(1 + r_i)(r_i(1 - \alpha_j) + \alpha_i) \;=\; 0 \;.$$

Two possibilities appear: $r_i = -1$ or $r_i = -\alpha_i/(1-\alpha_j)$.

For the first possibility, $r_1 = r_2 = -1$, Condition (A.9) is not satisfied. However, Equation (A.8) reduces here to:

$$-\alpha_i \ (e_i + e_j) \ = \ 0 \ ,$$

so that all strategy profiles such that $e_i + e_j = 0$ form a conjectural equilibrium. Since $V^i(e_i, e_j) = 0$ for such values, we are in the situation of Theorem A.2: conjectured reaction functions are inverse of one another, and are also iso-utility functions of both agents. However, from an economic standpoint, this solution makes little sense: it assumes a negative contribution from one agent, and yields a null utility.

The second consistent constant CVE is given by:

$$r_i^c \ = \ -\frac{\alpha_i}{1-\alpha_j} \ , \qquad e_i^c \ = \ (1-\alpha_i)I_i - \alpha_i I_j \ ,$$

it corresponds to a conjectured reaction function

$$\rho_j^c(e_i) \ = \ r_j^c \ (e_i - e_i^c) \ + \ e_j^c \ .$$

For these values of $r_i = -\frac{\alpha_i}{1-\alpha_j}$, Condition (A.9) holds under the assumption $\alpha_1 + \alpha_2 < 1$. The utility obtained by agent i at the equilibrium is:

$$V^i(e_i^c, e_j^c) \ = \ (I_i + I_j) \ \alpha_i^{\alpha_i} \ (1 - \alpha_i - \alpha_j)^{1-\alpha_i} \ .$$

Since this is strictly positive, this strategy profile Pareto-dominates any one obtained in the case $r_1 = r_2 = -1$. It is legitimate to consider it as the only constant CCVE of the game.

Observe that in this case, the conjectures do not satisfy $\rho_i^c \circ \rho_j^c = \text{id}$: unlike the first possibility $r_1 = r_2 = -1$ (and also the Cournot duopoly model of Section C.1), the conjectured reaction functions are not the inverse of one another.

The equilibrium is indeed a maximum for both possibilities of CCVE: we have here $\mu_i + \nu_i = \alpha_i + (1-\alpha_i) = 1$, and it has been observed on page 130 that this is a sufficient condition for maximality.

The sets of conjecture curves, iso-utility curves and conjectured reaction functions are displayed in Figure A.1 on page 122. The parameters used are $I_1 = I_2 = 1$, $\alpha_1 = 0.2$ and $\alpha_2 = 0.4$.

Voluntary contributions to a public good for many players

Now reconsider the previous model with many players. The individual utility function has the form:

$$V^i(e_i, E_{-i}) = (I - e_i)^\alpha (e_i + E_{-i})^{1-\alpha} ,$$

where $E_{-i} = \sum_{j \neq i} e_j$. The first order condition for a constant conjectural variations equilibrium gives

$$-\alpha(e_i + E_{-i}) + (1-\alpha)(1+r)(I - e_i) = 0 ,$$

so that the symmetric equilibrium is given by:

$$e^c = \frac{I(1-\alpha)(1+r)}{(1-\alpha)(1+r) + n\alpha} .$$

The second order condition for maximisation is

$$\begin{aligned} &- \big((I-e_i)^{\alpha-2}(e_i + E_{-i})^{1-\alpha} + 2(I-e_i)^{\alpha-1}(e_i+E_{-i})^{-\alpha}(1+r) \\ &\quad +(I-e_i)^{\alpha}(e_i+E_{-i})^{-\alpha-1}(1+r)^2\big) \ < \ 0 , \end{aligned}$$

which is always true. Therefore, $r > -1$ ensures the existence of a CVE $e^c > 0$. Moreover,

$$\begin{aligned} &V^i_{ii}(e_i, E_{-i}) + (n-1)V^i_{ij}(e_i, E_{-i}) \\ &\quad = -\alpha(1-\alpha)(I-e_i)^{\alpha-2}(e_i+E_{-i})^{-1-\alpha}(I+E_{-i})(E_{-i} + n(I-e_i) + e_i) \end{aligned}$$

is negative and (**H1**) holds. It is easily checked that Hypothesis (**H2**) holds, just as in the two-player case.

The following results are readily obtained:

- If the conjectural variations are of the form $r = e_i/E_{-i}$ or $r = e_i/(E_{-i} + e_i)$ *i.e.* $r = 1/(n-1)$ or $r = 1/n$ in the symmetric case (as in Cornes and Sandler (1984)), Theorem B.1 Case 3 indicates that the constant CVE Pareto-improves over the Nash equilibrium.
- As far as CCVE are concerned, Sugden (1985) concludes that there do not exist positive constant consistent conjectural variations for the model of voluntary contributions to a public good with n agents (for the Cobb-Douglas utility function). Therefore, if a constant consistent CVE exists, it must be negative; so Case 3 of Theorem B.1 holds and a CCVE would produce a Pareto-worsening over the Nash equilibrium.

C.3 A model of competition between regions

This is a model of competition between regions due to Wildasin (1991), already evoked in Chapter 2.

In this example, we will show that, under adequate assumptions, (**H1**) and (**H2**) are satisfied. Furthermore, in a quadratic formulation of the model, there exist two consistent conjectural variations equilibria, and one of them gives a Pareto-improvement over the Nash equilibrium.

Two regions $i = 1, 2$, are located around the same watershed and get benefit from it. The representative agent in region i has a utility function $U^i(x_i, s_i) = x_i + s_i$, where x_i is a private consumption good and s_i an index of the water quality. The production for the water quality is given by a decreasing return to scale technology. Formally $s_i = f(e_i, e_j)$ where $f(.,.)$ is a continuous and globally concave function and $e_i \in [0, \overline{E}]$ are the public expenses in region i. Region i is endowed with an exogenous revenue y_i, which is used to purchase the private good and to contribute to the water quality; therefore the budget constraint is $x_i + e_i = y_i$. Plugging $f(e_i, e_j)$ and the budget constraint in the utility function $U^i(x_i, s_i)$, the problem of the representative agent in region i is:

$$\max_{e_i} V^i(e_i, e_j), \text{ with } \quad V^i(e_i, e_j) = y_i - e_i + f(e_i, e_j) \ .$$

Given that the strategy spaces are compact and convex, the existence of a Nash equilibrium follows as the function $f(.,.)$ is continuous and $f_{11}(e_i, e_j) \leq 0$ (here this inequality is actually a consequence of the assumption of decreasing returns to scale); also there exists a symmetric Nash equilibrium, since the game is symmetric. To rule out corner solutions it is further imposed that:

$$\lim_{e_i \to 0} V_i^i(e_i, .) = -1 + f_1(e_i, .) > 0, \text{ and } \lim_{e_i \to \overline{E}} -1 + f_1(e_i, .) < 0 \ , \quad i = 1, 2.$$

Finally, Inequality (B.4) of assumption (**H1**) reads as

$$f_{11}(e_i, e_j) + f_{12}(e_i, e_j) \ \leq \ 0 \ .$$

This inequality necessarily holds when the technology is such that $f_{12}(e_i, e_j) \leq 0$. On the other hand if $f_{12}(e_i, e_j) \geq 0$, both (B.4) and (B.5) may obtain. But for (B.5) to hold, this is a necessary condition. As for the uniqueness of the Nash equilibrium and for the concept of consistent conjectural variations equilibria, very little can be said outside specific examples.

Accordingly we shall consider a quadratic formulation,

$$V^i(e_i, e_j) = a_0 + a_1\, e_i + a_2\, e_j + \frac{a_3}{2}\, e_i^2 + a_4\, e_i e_j + \frac{a_5}{2}\, e_j^2 \ ,$$

with $a_0 \geq 0, a_1 > 0,\ a_3 < 0,\ a_4 > 0$ and $a_5 \leq 0$.[5] Then we have $V_{ij} > 0$, $V_{ii} < 0,\ V_{jj} < 0$. The marginal externality is: $V_j(e_i, e_j) = a_2 + a_4 e_i + a_5 e_j$. Under mild conditions on the model parameters, (**H1**) and (**H2**) are satisfied. This is the case for instance if $2a_4 < -a_3$, $0 < a_1$, and $\frac{2-a_1}{a_4+a_3} < \overline{E}$. To characterise the conjectural equilibrium with a constant conjecture r let us compute the first order condition:

$$a_1 + ra_2 + a_3 e_i + a_4 e_j + a_4 r e_i + a_5 r e_j \ = \ 0 \ ,$$

given that $e_j = \rho_j^c(e_i)$ with $(\rho_j^c)' = r$, then the second order condition:

$$a_3 + 2ra_4 + r^2 a_5 \ > \ 0 \ .$$

So, if for instance, $a_4 \leq \sqrt{a_3 a_5}$, there is a maximum at:

$$e^c \ = \ - \ \frac{a_1 + ra_2}{(a_3 + a_4) + r(a_4 + a_5)} \ ,$$

which is the conjectural equilibrium with a constant conjecture r.
As $V_{ij} > 0$, from Section B.3 every consistent constant conjectural variations, if any exists, must be positive. It is easy to see that the CCVE must be:

$$r \ = \ \frac{1}{2a_4} \left(-(a_3 + a_5) \pm \sqrt{(a_3 + a_5)^2 - 4a_4^2} \right) \ . \tag{C.3}$$

Given the signs assumed for the parameters, consistent conjectural variations exist when $a_3 + a_5 \leq -2a_4$. If $a_3 + a_5 = -2a_4$, $r = 1$ and the Pareto solution results. If the inequality is strict, there exist two distinct and positive consistent conjectural variations: one of them is larger than 1, and the other one is smaller. Whatever the sign of the externality, the conjectural equilibrium for the conjecture $0 < r < 1$ gives a higher utility level than the Nash equilibrium; agents under-internalise the externalities. With the conjecture $r > 1$, agents over-internalise the externalities and their behaviours might result in a lower utility level than the one of the Nash equilibrium. When $V_i(e_i, e_j) > 0$ (public underspending), Theorem B.1, Case 1 holds. Should $V_i(e_i, e_j)$ be negative (public overspending), then Theorem B.1, Case 2 holds.

[5] In particular, $a_4 > 0$ captures the case of strategic complementarity; strategic substitutability would be modelled when $a_4 < 0$.

Finally, if $a_4 > \sqrt{a_3 a_5}$, the first order condition yields a maximum of the payoff when

$$r < -\frac{a_4 - \sqrt{a_4^2 - a_3 a_5}}{a_5} \text{ or } r > -\frac{a_4 + \sqrt{a_4^2 - a_3 a_5}}{a_5} .$$

In order to know if constant consistent CVE exist, it must be verified if r given by (C.3) belongs to one of these sets.

C.4 A model of aggregate-demand externalities

This model, proposed by Diamond (1982), is a macroeconomic model with aggregate demand externalities where assumption (**H1'**) and (**H2**) are simultaneously satisfied. In a quadratic version of the model, constant CVE exist for positive and negative values of r, but there are no constant *consistent* CVE and an interior Nash equilibrium at the same time. The setup can be presented within the restrictive interpretation of a search model. Two symmetric agents have payoff functions

$$V^i(e_i, e_j) = \alpha\, e_i\, e_j - C(e_i) ,$$

where e_i is player i's search intensity, $e_i e_j$ is the probability of finding a trading partner (thus, $E = [0,1]$), $\alpha > 0$ is the gain when a partner is found, and $C(e_i)$ is a continuous and convex cost function of the search effort.

The payoff functions are symmetric and concave with respect to the strategy of its associated player. The externality is $V_j^i = \alpha e_i > 0$. Here Inequality (B.5) in (**H1'**) holds when the second order derivative of the cost function is bounded by α:

$$\alpha \geq C''(e_i) .$$

And for (**H2**) to be satisfied then:

$$(G^P)'(e) = 2\alpha - C''(e)$$

must not change its sign over the interval $E = [0,1]$.

Turning attention to the uniqueness problem for the Nash equilibrium, and to the existence of a consistent conjectural variations equilibrium, we shall assume that

$$C(e_i) = \frac{c}{2}e_i^2 + be_i ,$$

with the parameters b and c all strictly positive. A necessary (second order) condition for the existence of a CVE with a constant conjecture r (for both players) is $2\alpha r < c$; the CVE is then given by:

$$e^c = \frac{b}{\alpha(1+r) - c} .$$

This equilibrium lies in the interval $(0,1)$ if and only if $r > \frac{b+c}{\alpha} - 1$. To summarise, the CVE exists when $\frac{b+c}{\alpha} - 1 < r < c/2\alpha$. In particular for $r = 0$, there exists an interior Nash equilibrium when $b + c < \alpha$. Note that in this example there exists a Nash equilibrium at the boundary. Indeed, (1,1) is a Nash equilibrium which is also Pareto optimal.

Moreover, $V^i_{ii} + V^i_{jj} = \alpha - c$, so $b + c < \alpha$ also implies that Assumption (**H1'**) is satisfied. In this case (**H2**) is satisfied since G^P is affine.

To conclude, if $\frac{b+c}{\alpha} - 1 < r < 0$, then by Theorem B.2, Case 3, the conjectural equilibrium produces a Pareto-improvement as compared to the interior Nash equilibrium. Moreover in this example

$$e^N \le e^c \le 1, \quad V^i(e^N, e^N) \le V^i(e^c, e^c) \le V^i(1,1) .$$

In other words, the CVE produces a Pareto-worsening with respect to the boundary Nash equilibrium. If $0 < r < c/2\alpha$, by Theorem B.2, Case 1, one obtains:

$$e^c \le e^N \le 1, \quad V^i(e^c, e^c) \le V^i(e^N, e^N) \le V^i(1,1) .$$

The problem of locating the consistent CVE with respect to the Nash outcome is pointless here. Indeed, as we proceed to show, the condition for a CVE to be consistent is not compatible with the existence of interior Nash equilibria. The payoff functions are given by:

$$V^i(e_i, e_j) = \alpha\, e_i\, e_j - \left(\frac{c}{2}e_i^2 + be_i\right) = e_i \left(\alpha e_j - \frac{c}{2}e_i + b\right) .$$

This is a special case of the general form of Theorem A.6 with $a'_i = 1$, $a_i = 0$, $b_i = 0$, $c_i = -c/2$, $c'_i = \alpha$, $d_i = b$, and $\mu_i = \nu_i = 1$. Accordingly, the constant CCVE are the solutions of Equation (A.12), that is:

$$-\ c\alpha\, r_i^2 + c^2\, r_i - \alpha c = 0 .$$

If the condition

$$c^2\ \left(c^2 - 4\,\alpha^2\right) \ge 0 \iff -\frac{c}{2} \le \alpha \le \frac{c}{2}$$

is satisfied, there are two possible values for the conjectures:

$$r_i^{\pm} = \frac{c \pm \sqrt{c^2 - 4\,\alpha^2}}{2\alpha} .$$

These two values have a product equal to unity. Therefore, we have $0 < r^- \leq 1 \leq r^+$, and equality occurs when $c = \pm 2\alpha$. Given that $\alpha, c > 0$, the condition for the existence of consistent CVE is actually $c \geq 2\alpha$, and we have seen that the condition for the existence of an interior Nash equilibrium is $c < \alpha - b$. Since $b > 0$, these are incompatible conditions.

Bibliography

Başar, T. and Olsder, G.J. (1999), *Dynamic Noncooperative Game Theory*, Second Ed., SIAM, Philadelphia.

Başar, T., Turnovsky, S.J. and d'Orey, V. (1986), "Optimal Strategic Monetary Policies in Dynamic Interdependent Economies", in *Springer Lecture Notes on Economics and Management Science*, Vol. 265, pp. 134–178.

Binmore, K.G. (1992), "Foundations of Game Theory", *in* J.-J. Lafont, ed., *Advances in economic theory: Sixth World Congress.* Vol. 1. Econometric Society Monographs, no. 20, pp. 1–31, Cambridge University Press.

Blaug, M. (1992), *The Methodology of Economics: Or How Economists Explain*, 2nd ed. Cambridge University Press, Cambridge.

Board, O. (2002), "Knowledge, Beliefs, and Game Theoretic Solution Concepts", *Oxford Review of Economic Policy*, Vol. 18 (4), pp. 418–432.

Bordignon, M. (1989), "The Nash Contribution Model: Theoretical Results, Empirical Test and Proposed Solutions", *Rivista di diritto finanziario e scienza delle finanze*, Vol. 48, pp. 502–549.

Bordignon, M. (1994), "A Further Look at Consistent Conjectures and Private Provision of Public Goods", *Ricerche Economiche*, Vol. 48, pp. 109–121.

Bowley, A.L. (1924), *The Mathematical Groundwork of Economics*, Oxford University Press, Oxford.

Boyer, M. and Moreaux, M. (1983a), "Consistent versus Non-Consistent Conjectures in Duopoly Theory: Some Examples", *The Journal of Industrial Economics*, Vol. XXXII, pp. 97–110.

Boyer, M. and Moreaux, M. (1983b), "Conjectures, Rationality and Duopoly Theory", *Intl. Journal of Industrial Organization*, Vol. 1, pp. 23–41.

Bresnahan, T.F. (1981), "Duopoly Models with Consistent Conjectures", *American Economic Review*, Vol. 71 (5), pp. 934–945.

Bresnahan, T.F. (1983), "Existence of Consistent Conjectures: Reply", *American Economic Review*, Vol. 73, pp. 457–458.

Cabral, L.M.B. (1995), "Conjectural variations as a Reduced Form", *Economics Letters*, Vol. 49, pp. 397–402.

Cornes, R. and Sandler, T. (1984), "The Theory of Public Goods: Non Nash Behaviour", *Journal of Public Economics*, Vol. 23, pp. 367–379.

Cornes, R. and Sandler, T. (1996), *The Theory of Externalities, Public Goods and Club Goods*, 2nd ed., Cambridge University Press.

Diamond, P. (1982), "Aggregate Demand Management in Search Equilibrium", *Journal of Political Economy*, Vol. 89, pp. 881–894.

Dixon, H. and Somma, E. (2001), "The Evolution of Consistent Conjectures", Discussion Papers in Economics, N° 2001/16, University of York, forthcoming in *Journal of Economic Behaviours and Organization.*

Dockner, E.J. (1992), "A Dynamic Theory of Conjectural Variations", *The Journal of Industrial Economics*, Vol. XL (4), pp. 377–395.

Driskill, R.A. and McCafferty, S. (1989), "Dynamic Duopoly with Adjustment Costs: a Differential Game Approach", *Journal of Economic Theory*, Vol. 49, pp. 324–338.

Fershtman, C. and Kamien, M.I. (1985), "Conjectural Equilibrium and Strategy Spaces in Differential Games", *Opt. Control Theory and Economic Analysis*, Vol. 2, pp. 569–579.

Fershtman, C. and Nitzan, S. (1991), "Dynamic Voluntary Provision of Public Goods", *European Economic Review*, Vol. 35, pp. 1057–1067.

Figuières, C. (2000), "Dynamic Concepts of Interaction in Long Run Investment Games", GREQAM working paper 00A01, University of Aix-Marseille 2.

Figuières, C. (2002), "Complementarity, Substitutability and the Strategic Accumulation of Capital", *International Game Theory Review*, Vol. 4(4), pp. 371–390.

Friedman, J.W. (1968), "Reaction Functions and the Theory of Duopoly", *Review of Economic Studies*, pp. 257–272.

Friedman, J.W. (1977), *Oligopoly and the Theory of Games*, North-Holland, Amsterdam.

Friedman, J.W. (2003), Private communication.

Friedman, J.W. and Mezzetti, C. (2002), "Bounded Rationality, Dynamic Oligopoly, and Conjectural Variations", *Journal of Economic Behavior and Organization*, Vol. 49, pp. 287–306.

Frisch, R. (1933), "Monopole, Polypole – La notion de force en Économie", *Nationaløkonomisk Tidsskrift*, Vol. 71, pp. 241–259 (reprinted "Monopoly, Polypoly: The concept of force in the economy" *in International Economic Papers*, Vol. 1, pp. 23–36, 1951).

Fudenberg, D. and Levine, D.K. (1998), *The Theory of Learning in Games*, MIT Press.

Fudenberg, D. and Tirole, J. (1991), *Game Theory*. MIT Press, 1991.

Gibbons, R. (1992), *A Primer in Game Theory*, Pearson Higher Education.

Hands, D.W. (2001), *Reflection without Rules: Economic Methodology and Contemporary Science Theory*, Cambridge University Press.

Holt, C.A. Jr. (1980), "Equilibrium Models of Tacit Collusion in Oligopoly Experiments with Price Setting Firms", Discussion paper 80-138m, Dept. of Economics, Univ. of Minnesota.

Horn, R.A. and Johnson, C.R. (1985), *Matrix Analysis*, Cambridge University Press.

Horn, R.A. and Johnson, C.R. (1991), *Topics in Matrix Analysis*, Cambridge

University Press.

Itaya, J.-I. and Dasgupta, D. (1995), "Dynamics, Consistent Conjectures, and Heterogeneous Agents in the Private Provision of Public Goods", *Public Finance/Finances Publiques*, Vol. 50(3), pp. 371–389.

Itaya, J.-I. and Okamura, M. (2003), "Conjectural Variations and Voluntary Public Good Provision in a Repeated Game Setting", *Journal of Public Economic Theory*, Vol. 5(1), pp. 51–66.

Itaya, J.-I. and Shimomura, K. (2001), "A Dynamic Conjectural Variations Model in the Private Provision of Public Goods: a Differential Game Approach", *Journal of Public Economics*, Vol. 81, pp. 153–172.

Jean-Marie, A. and Tidball, M. (2002), "Adapting Behaviors through a Learning Process", LAMETA Working paper DT 2002-10, University of Montpellier 1.

Kalai, E. and Stanford, W. (1985), "Conjectural Variations Strategies in Accelerated Cournot Games", *Intl. Journal of Industrial Organization*, Vol. 3, pp. 133–152.

Kamien, M. and Schwartz, N. (1983), "Conjectural Variation", *Canadian Journal of Economics*, pp. 192–211.

Laffont, J.-J. (1988), *Fundamentals of Public Economics*, MIT Press.

Laitner, J. (1980), "Rational Duopoly Equilibria", *The Quarterly Journal of Economics*, pp. 641–662.

Lindh, T. (1992), "The Inconsistency of Consistent Conjectures. Coming back to Cournot", *Journal of Economic behaviour and Optimization*, Vol. 18, pp. 69–90.

Müller, W. and Normann, H.-T. (2002), "Conjectural Variations and Evolutionary Stability: a new Rationale for Consistency", manuscript.

Olsder, G.J. (1981), "A Critical Analysis of a New Equilibrium Concept", Memorandum Nr. 329, Dept. Applied Maths., Twente University of Technology, The Netherlands.

Osborne, M.J. and Rubinstein, A. (1994), *A Course in Game Theory*, MIT Press.

Perry, M. (1982), "Oligopoly and Consistent Conjectural Variations", *Bell Journal of Economics*, Vol. 13, pp. 197–205.

Robson, A.J. (1983), "Existence of Consistent Conjectures: Comment", *American Economic Review*, Vol. 73, pp. 454–456.

Slade, M. (1995), "Empirical Games: The Oligopoly Case", *Revue canadienne d'Économique*, XXVIII (2).

Sterdyniak, H. and Villa, P. (1993), "Équilibres Conjecturaux Cohérents et Coordination de Politiques Économiques", *Revue d'Économie Politique*, Vol. 103 (2), pp. 221–251 (in French).

Sugden, R. (1982), "On the Economics of Philanthropy", *The Economic Journal*, Vol. 92, pp. 341–350.

Sugden, R. (1985), "Consistent Conjectures and Voluntary Contributions to Public Goods: Why the Conventional Theory does not Work", *Journal of Public Economics*, Vol. 27, pp. 117–124.

Tirole, J. (1988), *The Theory of Industrial Organization*, The MIT Press.

Ulph, D. (1983), "Rational Conjectures in the Theory of Oligopoly", *Intl. Journal of Industrial Organization*, Vol. 1, pp. 131–154.

van der Weel, H. (1975), *Endogene Prijstellingstheorie*, Stenfert Kroese, Leiden, The Netherlands (in Dutch).

Weibull, J.W. (1995), *Evolutionary Game Theory*, MIT Press, Cambridge, Mass.

Wildasin, D.E. (1991), "Some Rudimentary 'Duopolity Theory' ", *Regional Science and Urban Economics*, Vol. 21, pp. 393–421.

Index